TEACH YOURSELF BOOKS

ALGEBRA

This book provides an introduction to the principles and foundations of Algebra, with particular emphasis on its applications to engineering and allied sciences. The progressive exercises are designed both to test the reader's understanding of the subject and to give him practice in the essential power of manipulation. Only a knowledge of Arithmetic is assumed, though some reference is made to theorems in Geometry and Trigonometry for the benefit of those students who are acquainted with them.

Granted that it is possible to learn the subject from a book, then this volume will serve the purpose excellently. It covers the ground from the very beginning, along the usual paths of equations, factors, indices and so on, to a concluding chapter on simple progressions and an appendix which contains notes on Permutations, the Binomial Theorem and Quadratic Equations theory. A chapter worthy of especial mention is that dealing with Determination of Graph Laws, a topic too often neglected but one of great value and frequent practical use.

Regarded as a textbook, this is probably the best value for money on the market.

Higher Education Journal

TEACH YOURSELF BOOKS

ALGEBRA

P. Abbott, B.A.

ST. PAUL'S HOUSE WARWICK LANE
LONDON EC4P 4AH

First printed 1942
Revised edition 1971
Third impression 1974

ISBN 0 340 05501 4

Printed in Great Britain
for The English Universities Press, Ltd.,
by Richard Clay (The Chaucer Press), Ltd., Bungay, Suffolk

INTRODUCTION

ALGEBRA is such a wide and comprehensive subject that this volume cannot be regarded as anything more than an elementary introduction to it. It is an endeavour to enable the private student to learn something of the principles and foundations of the subject, thus enabling him to proceed to the study of more detailed and advanced treatises. It also provides, within the necessarily prescribed limits of such a book, that knowledge of Algebra which is required by a student of allied branches of Mathematics or in applications of Mathematics to Engineering, etc. Consequently some of those elementary sections of the subject which are of little use for these purposes have not been included.

The exercises are progressive and designed both to enable the student to test his knowledge of the work he has studied and also to provide material for his training in that power of manipulation which is so essential. They contain few of the more complicated or academic problems which are beyond the practical requirements of the ordinary student.

An Appendix contains, without exercises, a very brief summary of the meaning of Permutations and Combinations, the Binomial Theorem, and the nature of the roots of a Quadratic Equation, together with those formulae which students may require when beginning work on the Calculus or other branches of Mathematics.

While the fundamental laws of Algebra have not been entirely overlooked, rigid proofs of them have been omitted, owing to exigencies of space. It is hoped, however, that the logical basis of the subject has not been seriously impaired by the omissions.

Some emphasis has been placed on the graphical aspects of parts of the subject, since experience has shown that they prove stimulating and provide revealing help to the student.

No previous mathematical knowledge is required for this work, beyond that of Arithmetic. References have occasionally been made to theorems in Geometry or Trigonometry for the benefit of those students who have some knowledge of them.

The Author is desirous of expressing his indebtedness to Mr. C. E. Kerridge, B.Sc., for the use of a number of examples from *National Certificate Mathematics*, Vol. I, and also to Mr. H. Marshall, B.Sc., for the use of examples from Vol. II of the same work. He also desires to record his gratitude to Mr. S. R. Morrell for the valuable assistance he has given in the correction of proofs.

P. ABBOTT.

PUBLISHER'S NOTE TO 1971 EDITION

This edition has been revised to include S.I. units.

CONTENTS

CHAPTER I
THE MEANING OF ALGEBRA

CHAPTER II
ELEMENTARY OPERATIONS

CHAPTER III
BRACKETS AND OPERATIONS WITH THEM

CHAPTER IV
POSITIVE AND NEGATIVE NUMBERS

CHAPTER V
SIMPLE EQUATIONS

CHAPTER VI
FORMULAE

CHAPTER VII
SIMULTANEOUS EQUATIONS

CONTENTS

CHAPTER VIII
GRAPHICAL REPRESENTATION OF QUANTITIES

CHAPTER IX
THE LAW OF A STRAIGHT LINE; CO-ORDINATES

CHAPTER X
MULTIPLICATION OF ALGEBRAICAL EXPRESSIONS

CHAPTER XI
FACTORS

CHAPTER XII
FRACTIONS

CHAPTER XIII
GRAPHS OF QUADRATIC FUNCTIONS

CHAPTER XIV
QUADRATIC EQUATIONS

CONTENTS

CHAPTER XV
INDICES

CHAPTER XVI
LOGARITHMS

CHAPTER XVII
RATIO AND PROPORTION

CHAPTER XVIII
VARIATION

CHAPTER XIX
THE DETERMINATION OF LAWS

CHAPTER XX
RATIONAL AND IRRATIONAL NUMBERS; SURDS

CHAPTER XXI
ARITHMETICAL AND GEOMETRICAL SERIES

APPENDIX

CHAPTER I

THE MEANING OF ALGEBRA

1. Algebra and Arithmetic.

ALGEBRA, like Arithmetic, deals with numbers. Fundamentally the two subjects have much in common; indeed, Algebra has been called " generalised Arithmetic ", though this is a very incomplete description of it. It would perhaps be more correct to say that Algebra is an extension of Arithmetic.

Both subjects employ the fundamental operations of addition, subtraction, multiplication and division of numbers, subject to the same laws. In each the same symbols, $+$, $-$, $\times$, $\div$, are used to indicate these operations, but in Algebra, as new processes are developed, new symbols are invented to assist the operations. Terms such as fractions, ratio, proportion, square root, etc., have the same meaning in both subjects, and the same rules govern their use.

In Arithmetic we employ definite numbers; we operate with these and obtain definite numerical results. Whereas in Algebra, while we may use definite numbers on occasions, we are, in the main, concerned with general expression and general results, in which letters or other symbols represent numbers not named or specified.

This may seem vague to a beginner, but the following simple example may serve to show what is meant.

2. A formula.

In Arithmetic we learn that to find the area of the floor of a room, rectangular in shape, we " multiply the length of the room by the breadth ". This might be expressed in the form :

The area of a rectangle in square metres is equal to the length in metres multiplied by the breadth in metres.

This rule is shortened in Algebra by employing letters as symbols, to represent the quantities. Thus:

Let the letter l represent " the length in metres ".

    „   „ b „   " the breadth in metres ".

    „   „ A „   " the area in square metres ".

With these symbols the above rule can now be written in the form :

$$A = l \times b.$$

In this shortened form we express the rule for finding the area of **any** rectangle; it is a general rule, and is called a **formula**. The full description of it would be

" *The formula for finding the area of a rectangle* ".

It will be noticed that in the formula above there is no mention of units. This is because it is true whatever units are employed. It is necessary, however, that the same kind of unit should be employed throughout. If l and b are measured in mm, A will be the area in square mm; if they are measured in m, A will be in square m.

When using the formula for any specific case it is important to state clearly what unit is employed for l and b. The unit for A will then follow.

Algebra has been called a kind of shorthand, and the above example in which a sentence has been reduced to

$$A = l \times b$$

illustrates the reason for the description, but as progress is made the student will discover that it is an incomplete description. From such a simple beginning as the above the subject develops into the most powerful instrument employed in Mathematics.

The Greeks produced some of the greatest mathematicians in history, but their work was mainly accomplished in Geometry. They made very little progress in Algebra,. since the symbols they could employ were extremely few. They did not even possess separate symbols for numbers, such as were afterwards introduced through the Arabs, and which we use to-day, but used instead the letters of the alphabet to represent numbers. The Romans were similarly restricted, and neither nation employed the decimal system of notation. The student will realise something of the great difference suitable symbols make in Mathematics if he will, as an example, write down the number of the year in Roman numerals and then try to multiply it, say by 18, also expressed in Roman fashion.

For progress in manipulation and general development of the subject not only do we require symbols, but they must be suitable for the purpose. The history of Mathematics reveals that many of the symbols which are so familiar to us now have been reached by a slow evolutionary process, often lasting through centuries.

When letters are used to represent numbers any suitable choice may be made. In the above formula, although A, l and b were employed, any other letters could be used if they were more suitable. By common usage, however, certain letters are usually selected for specific purposes, and the same symbols are used invariably to denote certain numbers, as will become apparent to the student as he progresses.

But whatever letters are employed in solving problems, the student must carefully observe the following:

(1) It must be clearly stated what each symbol represents.
(2) If measurements of any kind are involved, the unit employed must be clearly defined.

3. Transformation of a formula.

The student may be inclined to wonder why such an elementary rule as the area of a rectangle should be expressed as a formula by the employment of algebraic symbols. But this is an example selected for its simplicity to illustrate the meaning of an algebraic formula. Even in this case it is possible in some measure to illustrate the flexibility and adaptability of a formula as compared with a statement in words.

For example, suppose that l and b were measured in cm, then $A = l \times b$ square centimetres.

If it is required to express this result in square metres we could do it thus:

$$A = \frac{l \times b}{10\ 000} \text{ square metres}$$

Suppose that the rectangle is a room, l and b being measured in cm, and that the room is covered with carpet costing a pence per square metre. If the cost of the whole carpet be represented by C pence, then:

Since $$A = \frac{l \times b}{10\ 000} \text{ square metres}$$

then $$C = \frac{l \times b}{10\ 000} \times a \text{ pence.}$$

If this were to be expressed in pounds, then:

$$C = £\frac{l \times b}{10\ 000} \times \frac{a}{100}.$$

Thus a formula may be used as a foundation for other formulae to express modifications of the original.

There is a convenient notation which is employed throughout mathematics and it is more fully discussed in §18. Briefly, when a number is multiplied by itself, this is called *squaring* the number: thus, $a \times a$ is a squared and is written a^2. Similarly, 1 square metre, the unit of area which is 1 m × 1 m, is written as 1 m². 28 square metres would be written 28 m². Again, m³ is m × m × m and represents cubic metres.

4. An illustration from numbers.

The following example illustrates the use of letters to represent generalisations in number. We know that:

(1) If any integer—*i.e.*, a whole number—be multiplied by 2, the result is always an **even number**.

(2) If any even number be increased by unity, the result is an **odd number**.

These two statements can be combined in one as follows:

If any integer be multiplied by two and the product increased by unity, the result is an odd number.

This is a generalisation about an odd number, expressed

in words. This can be expressed by means of algebraical symbols as follows:

(1) Let n be any integer.
(2) Then $2 \times n$ is always an **even** number.
(3) Therefore $2 \times n + 1$ is always an **odd** number.

In (3) we have reached an algebraical expression by means of which any odd number can be represented.

The brevity and lucidity of this expression as compared with the full description of an odd number above will be apparent. But its value goes beyond this. We can manipulate this algebraic form, we can operate with it, and so can use it in the solution of problems.

We may first note, however, that when expressing the product of two or more numbers represented by letters or a numeral and letters, the sign of multiplication can be omitted. Thus $2 \times n$ can be written as $2n$, and $2 \times n + 1$ as $2n + 1$. This cannot be done with two numerals, such as 25, because under the decimal system the figure 2 has a place value. Multiplication may also be shown by a dot, thus $2 \cdot n$.

If any odd number can be represented by $2n + 1$, then, since when any odd or even number is increased by 2 the result is the next odd or even number, therefore the next odd number greater than $2n + 1$ is $2n + 1 + 2$ or $2n + 3$. Similarly $2n + 5$ is the next odd number above $2n + 3$. Consequently the expression

$$2n + 1, 2n + 3, 2n + 5, 2n + 7 \ldots \text{etc.,}$$

represent a series, or a succession of consecutive **increasing odd numbers.**

Similarly, if an even number be diminished by 1, we obtain an odd number.

$\therefore 2n - 1, 2n - 3, 2n - 5 \ldots$, represent a series of **decreasing odd numbers.**

Note.—The succession of "dots" after the sets of odd and even numbers indicates that we could write down more such numbers if it were necessary.

5. Substitution.

In the algebraic representation of a set of odd numbers—viz.:

$$2n + 1, 2n + 3, 2n + 5 \ldots$$

since n represents **any** integer, we could, by assigning some particular value to n, obtain the corresponding odd number.

Thus if $\qquad n = 50.$

Then $\qquad 2n + 1 = 2 \times 50 + 1$
$$= 100 + 1$$
$$= 101.$$

Similarly, $\qquad 2n + 3 = 2 \times 50 + 3$
$$= 100 + 3$$
$$= 103.$$

Consequently the series of increasing odd numbers corresponding to this particular value of n is:

$$101, 103, 105, 107 \ldots$$

Similarly, the decreasing odd numbers when $n = 50$ can be found by substituting this value of n in the series:

$$2n - 1, 2n - 3, 2n - 5 \ldots$$

Then we get the corresponding arithmetical values:

$$99, 97, 95 \ldots$$

6. Letters represent numbers, not quantities.

Such things as length, weight, cost, are called **quantities**. In short, anything which can be measured is called a quantity.

Letters are not used to represent these quantities. Thus in the formula which we stated above—viz.:

$$A = l \times b.$$

l stands for the **number** of cm or m, as the case may be, and to obtain it we must first determine the unit to measure the quantity, and then use a letter to represent the number of units.

It is important, therefore, that when letters are to be used in algebraical expressions, it should be clearly stated what each letter represents.

Thus we should say:

Let l represent the number of cm in the length or, more briefly, let the length be l cm.

Similarly let n represent the number of men, and let c represent the cost in pence.

We do sometimes, for brevity, and somewhat loosely, speak, for example, of the area of a rectangle as A, when we should say A m² or A units of area.

The sign $=$ may be used loosely in abbreviation of the statements above. Thus we could write:

Let $l =$ the length in metres,
or let $c =$ the cost in pence.

The sign $=$ means " equals " or " is equal to ". It should connect two expressions which are equal in magnitude, but is often used loosely as above to express equality.

7. Examples of algebraic forms.

We now give a few examples of what may be termed algebraic forms—*i.e.*, the expression in algebraic symbols and signs of operation of statements about quantities.

Example 1. *Express in algebraic form the number of pence in x pounds added to y pence.*

To express pounds in pence we multiply by 100.

∴ x pounds $= 100 \times x$ pence,
and the total number of pence is $100x + y$.

Example 2. *A car travels for t hours at v km/h. How far does it go? How far will it go in 20 min?*

The car goes v km in 1 h and ∴ $2v$ km in 2 h, $3v$ km in 3 h and so on; in t h it will go $t \times v$ km.

20 min is $\frac{1}{3}$ h and the car ∴ travels $\frac{1}{3} \times v$ km in 20 min.

In $\frac{1}{3}$ h the car travels $\dfrac{v}{3}$ kilometres.

Example 3. *There are two numbers; the first is multiplied by 3 and 5 is added to the product. This sum is divided by 4 times the second number. Express the result in algebraic form.*

We must begin by choosing letters to represent the unknown number.

Let x represent the first number.

   ,, y represent the second number.

Then three times x increased by 5 will be represented by $3x + 5$, and four times the second number is $4y$.

The division of the first expression by this is $\dfrac{3x + 5}{4y}$.

Exercise I.

1. Write down expressions for:

    (1) The number of pence in £x.
    (2)   ,,   ,,   pounds in n pence.

2. If £a be divided among n boys, how many pence will each boy get?

3. If n men subscribe £a each and m other men subscribe b pence each, how many pence are subscribed in all?

4. Write down the number of

    (1) Metres in a kilometres.
    (2) Men in x Apollo spacecraft.
    (3) Tonnes in y kilogrammes.

5. The sum of two numbers is 28. If one number be n, what is the other?

6. The difference between two numbers is x; if one of them is 50, what is the other?

7. The product of two numbers is a and one of them is x; what is the other?

8. If the average length of a man's step is x cm,

    (1) How far will he walk in 100 steps?
    (2) How many steps will he take in walking 1·8 km?

9. A number (n) is multiplied by 2, and 5 is added to the sum. Write down an expression for the result.

10. If x is an odd number write down expressions for

    (1) The next odd numbers above and below it.
    (2) The next even numbers above and below it.

11. A man buys sheep; x of them cost him a pence each and y of them cost b pence each. What was the total cost in pounds?

12. A number is represented by x; double it, add 5 to the result, and then divide the whole by $6y$. Write down an expression for the result.

13. What number must be subtracted from a to give b?

14. What number divided by x gives y as a quotient?

15. What is the number which exceeds b by a?

16. The numerator of a fraction is x increased by 2. The denominator is y diminished by 5. Write down the fraction.

17. A car travels for m hours at v km/h. It then travels n hours at u km/h. How far does it travel in all?

18. What is the total number of pence in

$$£a + b \text{ pence.}$$

19. A train travels at v km/h. How far does it go in x hours and how long does it take to go y km?

20. Two numbers, less than 10, are chosen. Add them; multiply this sum by 2; add 4; multiply by 3; add 4 times one of the original numbers; take away 12; take away 5 times the second of the original numbers. The result is $10m + n$. What were the original numbers? Interpret this result as a party trick.

CHAPTER II

ELEMENTARY OPERATIONS IN ALGEBRA

8. Use of symbols.

IN order that the student may become familiar with the processes of Algebra he needs considerable practice in the use of symbols. Consequently in this and subsequent chapters he will constantly be using letters which represent numbers in a general way and without any reference to quantities such as length, cost, etc., as was done in the previous chapter.

Thus when we employ the form $a + b$ we shall, in general, be using the letters, not as referring to any particular quantity, but as standing for any numbers.

9. Symbols of operation.

As stated in §1, certain symbols of operation, such as $+, -, \times, \div, \sqrt{}$, are common to Arithmetic and Algebra, since they are used for operations which are performed in both subjects. Usually, however, there is a certain difference in the way they are employed. It is evident that while such operations as

$$5 + 7,\ 10 - 3,\ 6 \times 4,\ 15 \div 3,\ \sqrt{9}$$

can be, and usually are carried out at once with definite numerical results, expressions such as

$$a + b,\ a - b,\ a \times b,\ a \div b,\ \sqrt{a}$$

cannot be evaluated numerically while a and b represent any numbers. Until numerical values are assigned to them we cannot proceed further with the operation. But we can, and do, operate with the expressions themselves, without any reference to their numerical values.

In addition to the above, many other symbols of operation are used in Algebra, among them the following:

Symbol.	Meaning.
=	See § 6.
≠	is not equal to.
≈	is approximately equal to.
>	is greater than.
<	is less than.

10. Algebraic expression. Terms.

Such a combination of letters and symbols as $2a + b$ is an example of what is called an **algebraic expression**. It may be defined as follows:

An algebraic expression is a combination of symbols which stand for numbers and for operations with them.

For brevity the term "expression" is usually employed. When the expression contains the symbols of operation $+$ or $-$, those parts of the expression which they separate are called **terms**.

Thus $2a + 3b$ is an expression of two terms or a **binomial**.

$\dfrac{5x}{2} - \dfrac{3y}{5} + 6z$ is an expression of three terms or a **trinomial**.

A combination of letters which does not contain either of the signs $+$ or $-$ may be said to be an expression of one term, or a **monomial**.

Thus $\dfrac{5ab}{6}$ is such an expression.

11. Brackets.

It frequently happens that an expression, or part of an expression, is to be operated as a whole. For example, suppose we wish to write in algebraic symbols " Twice the sum of a and b ".

Evidently the arrangement $2 \times a + b$ does not make it clear whether the 2 is to multiply a only or the sum of a and b. Consequently we employ " **brackets** " to enclose the part which is to be operated on as a whole—viz., $a + b$.

∴ we write $2(a + b)$.

In this arrangement the multiplication sign is omitted between the 2 and the bracket.

The brackets have the effect of indicating the **order** in which operations are to be carried out.   Thus:

$$2(a + b) - c$$

means that we find the sum of a and b, multiply this by 2 and then subtract c.

Similarly $(a + b) \times (c + d)$ or $(a + b)(c + d)$ means that we find the sum of a and b, and also of c and d and then multiply the two results.

This will be considered further at a later stage.

12. Coefficient.

The expression $3a$ denotes a multiple of a and the number 3, which indicates the multiple, is called the **coefficient** of a.

The coefficient may be a definite number like 3, called a numerical coefficient, or it may be a letter representing a number.

Thus in the expression ax, a may be regarded as the coefficient of x, but in some problems when we are thinking of multiples of a, x would be a coefficient of a; in such a case we would usually write the expression as xa.

In general, if an expression is the product of a number of factors, any one of them can be regarded as the coefficient of the product of the others, when for any purpose we regard this product as a separate number.

Thus in $3ab$,   3 is the coefficient of ab
$\qquad\qquad$ $3a$,,$\qquad$,,$\qquad$ b
$\qquad\qquad$ $3b$,,$\qquad$,,$\qquad$ a.

Like terms.   In an expression, terms which involve the same letter, and differ only in the coefficients of this letter, are called **like terms**.

Thus in the expression

$$3a + 5b - 2a + 4b$$
$$3a \text{ and } 2a \text{ are like terms.}$$
and $\qquad$ $5b$ and $4b$ $\qquad$,, $\qquad$,,

13. Addition and subtraction of like terms.

In Arithmetic we learn that the sum of

$$5 \text{ dozen and } 9 \text{ dozen is } 14 \text{ dozen,}$$
or $\qquad (5 \times 12) + (9 \times 12) = 14 \times 12$.   .   (A)

Similarly   8 score + 7 score = 15 score

or $(8 \times 20) + (7 \times 20) = 15 \times 20.$

So for any number as for example

$$(6 \times 24) + (11 \times 24) = 17 \times 24.$$

In Algebra, if we were to let a represent 12 in the statement (A) given at the beginning of this section, we could write

$$5a + 9a = 14a,$$

and for the other cases

$$8a + 7a = 15a$$

and $6a + 11a = 17a.$

These last three cases are generalised forms of the preceding examples, but it must be noted that whereas in the arithmetical forms we can proceed to calculate the actual value of the sum in each, in the algebraical forms we can proceed no further in the evaluation until a definite numerical value is assigned to a.

Subtraction leads to similar results, just as

9 dozen — 5 dozen = 4 dozen.

so $9a - 5a = 4a.$

In this way we can add or subtract like terms only. It is not possible, for example, to perform any addition of two unlike terms such as $9a + 5b$.

The rule for adding together like terms will now be clear. It is " add the coefficients ".

Thus the sum of $2x + 5x + 3x = 10x$, whatever x may be. The operation of "finding the sum" is used to include both addition and subtraction. This is called the " algebraic sum ".

When an expression contains more than one set of like terms, these should be collected and dealt with separately.

14. Worked examples.

Example 1.   *Simplify $5a + 6b + 2a - 3b$.*

Collecting like terms $5a + 2a = 7a$

$6b - 3b = 3b.$

Hence the whole expression is equal to $7a + 3b$.   In

practice there is generally no need to write down the above steps. The calculations can be made mentally.

Example 2. *Simplify* $15x - 3y + 6y + 7x - 5$.

Collecting like terms and adding coefficients we get

$$22x + 3y - 5.$$

15. The order of addition.

The counting of a number of things is not affected by the order in which they are counted. Thus :

$$6 \text{ apples} + 4 \text{ apples}$$

is the same in number as

$$4 \text{ apples} + 6 \text{ apples}.$$

This will be clear when it is remembered that 6 is the symbol for 6 units, and 4 is the symbol for 4 units. Thus :

$$6 + 4 = (1 + 1 + 1 + 1 + 1 + 1) + (1 + 1 + 1 + 1)$$

and

$$4 + 6 = (1 + 1 + 1 + 1) + (1 + 1 + 1 + 1 + 1 + 1).$$

In each case the total number of units is the same.

Thus algebraically $6a + 4a$ is the same in value as $4a + 6a$. This is true for any algebraical sum.

Thus $6a + 5b - 3a$ can be written as $5b - 3a + 6a$ without altering the value of the expression.

Briefly *the order in which numbers may be added is immaterial*.

16. Evaluation by substitution.

If we wish to find the numerical value of an algebraical expression for definite numerical values of the letters composing it, the expression should first be simplified by adding like terms. Then the numerical values are substituted for the letters.

Example. *Find the value of* $6x + 2y - 3x + 4y - 3$ *when* $x = 3$ *and* $y = 2$.

Simplify the expression as in § 14 :

$$6x + 2y - 3x + 4y - 3 = 3x + 6y - 3.$$

Substituting the given values :

$$3x + 6y - 3 = (3 \times 3) + (6 \times 2) - 3$$
$$= 9 + 12 - 3$$
$$= 18.$$

Note.—It will be seen that brackets are introduced when it is desirable to keep terms separate for evaluation.

Exercise 2.

1(a) Find the value of 6 dozen + 4 dozen.
  (b) Simplify $6a + 4a$ and find its value when $a = 12$.
2(a) Find the value of $(8 \times 73) - (3 \times 73)$.
  (b) Find the simplest form of $8b - 3b$ and find its value when $b = 73$.
3. Write down in its simplest form:

$$a + a + a + b + b + b + b$$

and find its value when $a = 5$ and $b = 8$.
4. Add together $2a$, $4a$, a, $5a$, and $7a$ and find the value of the sum when $a = 2 \cdot 5$.
5. Write the following expressions in their simplest forms:

  (1) $15b + 11b$.   (2) $15x - 3x + 7x$.
  (3) $9a - 4a + 6a + a$.   (4) $4x + 3x - 2x - x$.

6. Write in their simplest forms:

  (1) $5a - 2b - 3a + 6b$.   (2) $11p + 5q - 2q + p$.
  (3) $a - 2 + 3b + 6 + 5a$.

7. Add together:

  (1) $4a - 5b$, $a + 6b$, $5a + b$.
  (2) $b + c - 3d$, $c + 2b + d$, $d - b - c$.
  (3) $5x + 2y + 3z$, $x - y - 2z$, $2x - y + z$.

8. When $a = 2$, $b = 3$, find the numerical values of

  (1) $3a + 2b + 1$.   (2) $5a - 3b + 6$.
  (3) $6a + 2b - 3a + 1$.   (4) $4a - 5b - 2b + 12a$.

9. Simplify the following expressions and find their values when $a = 4$, $b = 2$, $x = 3$, $y = 5$.

(1) $4ab - 2ab + 6ab$.
(2) $5ax - 2ax + bx$.
(3) $6xy - 4xy + xy$.
(4) $ab + 6bx - ay + 3ab - 2bx$.

10. Find the numerical value of $a + \frac{1}{2}a + \frac{1}{4}a + \frac{1}{8}a$ when $a = 2$.

11. When $x = 1$, $y = 2$, $z = 3$, find the numerical value of

$$3x + 5y - 4z + 8x - y + 5z.$$

12. When $x = 4$, $y = 5$, $z = 1$, find the values of

(1) $3xy + 2yz - 3z - 1$.   (2) $xy + yz + zx$.
(3) $xy + y + x + 1$.

13. If n be an odd number, write down the next three odd numbers greater than it and find their sum.

14. Write down a series of four numbers of which the first is a, and each of the others is twice the one which precedes it. Find their sum.

15. Write down a series of 5 numbers of which the first is a, the second is greater than the first by d and each of the other three is greater by d than the one which precedes it. Find their sum.

16. There are 5 numbers, the smallest of which is expressed by $2n + 5$. Each of the others is 3 greater than the one which precedes it. Write down the numbers and find their sum.

---

17. Multiplication.

Order of multiplication. In Algebra, as in Arithmetic, the multiplication of a number of factors may be performed in any order, or, more precisely:

The product of a number of factors is independent of the order in which they are multiplied.

Thus 3×4 is equal in value to 4×3
        $6 \times 3 \times 5$,,   ,, $3 \times 5 \times 6$

and generally $\quad a \times b$ is equal in value to $b \times a$
and $\qquad a \times b \times c \qquad$,, $\qquad$,, $\qquad c \times b \times a$.

Consequently if it is required to multiply say, $2a$ by 5 we can write the product thus:

$$2 \times a \times 5 = 2 \times 5 \times a$$
$$= 10a$$
and $\qquad 3a \times 2b = 3 \times 2 \times a \times b$
$$= 6ab.$$

In this last example, a and b being unlike letters, we cannot proceed further with the multiplication.

It should be noted, however, that although 4×3 is equal in value to 3×4, the two products do not necessarily mean the same thing when they refer to quantities.

If, for example, 12 soldiers were to " form fours ", they would be arranged as shown in Fig. 1(a), the arrow showing the direction in which they are facing. But if the same 12 men

Fig 1(a)   Fig. 1(b).

were to " form threes " they would be arranged as in Fig. 1(b).

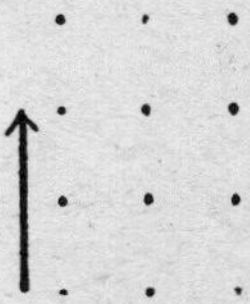

Thus 3 rows of 4 men require the same number of men as 4 rows of 3 men, but *they are a different arrangement*. Similarly if 4 men pay 3 pence each; the **total** amount paid is the same as that when 3 men pay 4 pence each.

18. Powers of numbers.

The product of equal numbers is called a power. Thus:

8×8 is called the **second** power of 8, or the **square** of 8.

$8 \times 8 \times 8$ is called the **third** power of 8, or the cube of 8, there being **three** equal factors.

$8 \times 8 \times 8 \times 8$ is called the **fourth** power of 8, there being **four** equal factors.

Similarly:

$\qquad a \times a$ is the second power of a or the **square** of a.
$\quad a \times a \times a \quad$,, $\quad$ third $\qquad$,, $\quad a \quad$,, $\quad$ **cube** of a.
$a \times a \times a \times a \quad$,, $\quad$ fourth $\qquad$,, $\quad a$.

The process of writing a power in full is tedious, and the form of it restricts further operations, especially when the power is a high one. Accordingly mathematicians made many attempts through centuries to devise a symbolic method of representing the row of factors. Finally, Descartes in 1637 used a numeral to mark the number of factors or the power and wrote the **cube** of a for example as a^3, the fourth power as a^4, etc.

The figure used in this way is called an **index** or exponent; it indicates the number of factors.

Thus $a \times a \times a \times a \times a$ is written as a^5. With this symbolic method it is as easy to write down the 20th power of a number as the 2nd.

But a new symbol, if it is to be satisfactory, must not only express clearly and concisely the purpose for which it was devised, but it must also be convenient for operations with it. We shall see, from what follows, that an index fulfils this condition, and later in the book it will be seen how it lends itself to important developments.

19. Multiplication of powers of a number.

Suppose we require to multiply two powers of a—say $a^2 \times a^3$. These numbers written in full are

$$(a \times a) \times (a \times a \times a)$$

the brackets serving to separate the two powers.

It is a fundamental law of Algebra, which will be assumed here, that when two groups of factors are multiplied, the factors in the groups are *associated* as one group of factors to give the product.

By this law $(a \times a) \times (a \times a \times a)$
         $= (a \times a \times a \times a \times a)$.

∴ *the number of factors in the product is the sum of the number of factors in the two groups.*

∴ in the example above the product $(a \times a \times a \times a \times a)$ is the 5th power of a, and **the index of the product is the sum of the indices of the two factors.**

∴ $a^2 \times a^3 = a^{2+3}$
                    $= a^5$.

The same reasoning may be applied to other cases and so

we may deduce the general rule for the multiplication of two powers of a number.

When two powers of the same number are multiplied, the index of the product is the sum of the indices of the factors.

Examples.

(1) $x^4 \times x^4 = x^{4+4} = x^8$.
(2) $2a^7 \times a^3 = 2a^{7+3} = 2a^{10}$.
(3) $5b^2 \times 3b^5 = 5 \times 3 \times b^2 \times b^5 = 15 \times b^{2+5} = 15b^7$.
(4) $a^2b \times ab^2 = a^2 \times a \times b \times b^2 = a^3b^3$.

The rule may be extended to the product of more than two factors. Thus:

$$x^2 \times x^3 \times x^4 = x^{2+3+4}$$
$$= x^9.$$

20. Power of a product.

To find the value of $(ab)^2$.

The use of the bracket shows that, as stated in § 11, the expression within the bracket must be regarded as a whole.

∴ by definition of a power

$$(ab)^2 = (ab) \times (ab)$$
$$= (a \times a) \times (b \times b)$$
$$= a^2b^2.$$

Thus we see that the effect of this is that the index 2 must be distributed over each of the factors.

Thus: $(2xy)^3 = 2^3 \times x^3 \times y^3$
$$= 8x^3y^3.$$

So in Arithmetic $(2 \times 5)^2 = 2^2 \times 5^2 = 4 \times 25 = 100$. Consequently *when taking a power of a product, the index of the power is said to be distributed and applied to each factor of the product.*

Exercise 3.

Write down the following in their simplest forms:

1. $4a \times 3$.
2. $5x \times 2y$.
3. $\frac{1}{2}x \times 4y$.
4. $7m \times 3n$.

5. $\frac{1}{3}a \times \frac{1}{4}b$.   6. $6a \times \frac{2}{3}b$.

7. $3a \times 4b \times 5c$.   8. $\frac{x}{2} \times \frac{y}{3} \times \frac{z}{4}$.

---

9. $x^2 \times x$.   10. $a \times a^2 \times a$.
11. $x^2 \times x^2$.   12. $a^3 \times a^3$.
13. $2a^2 \times a^3$.   14. $3x^3 \times 2x^4$.
15. $2ab \times ab$.   16. $2b \times 3b^3$.
17. $x^2y \times xy^2$.   18. $7x^3a \times x^3a$.
19. $2a \times 3a^2 \times a^3$.   20. $(3a^2b)^3$.
21. $(x^3)^2$.   22. $(2a^4)^3$.
23. $(2a^3)^4$.   24. $(4a)^2 \times 4a^2$.

Find the numerical values of the following:

25. $2a^2 \times a$, when $a = 3$.
26. $2a^2 + a$, when $a = 3$.
27. $a^2b \times ab^2$, when $a = 1$, $b = 2$.
28. $x^2 + 7x + 2$, when $x = 10$.
29. $3a^2 + 2ab + b^2$, when $a = 2$, $b = 3$.
30. $c^2 \times c^3 \times c^4$, when $c = 1$.
31. $3a \times 3a \times 3a$, when $a = 2$.
32. $(2a^2x)^2$, when $a = 2$, $x = 3$.

---

21. Division of powers.

Suppose that a power of a number is divided by another power of the same number, as for example,

$$a^5 \div a^2.$$

Every division can be expressed in fractional form, as in Arithmetic.

$$\therefore \qquad a^5 \div a^2 = \frac{a^5}{a^2}$$

$$= \frac{a \times a \times a \times a \times a}{a \times a}$$

(by definition of a power).

As in Arithmetic, common factors in the numerator and denominator can be cancelled.

∴ the two a factors in the denominator can be cancelled with two of the five factors of the numerator.

Then there will be left in the numerator $(5 - 2)$ factors.

$$\therefore \qquad a^5 \div a^2 = a^{5-2}$$
$$= a^3.$$

Clearly the same method can be followed whatever the powers. Consequently we may deduce the rule:

When dividing a power of a number by another power of the same number, subtract the index of the divisor from the index of the dividend.

Example. *Divide $84a^6$ by $12a^2$.*

$$\frac{84a^6}{12a^2} = 7a^{6-2} = 7a^4.$$

Example. *Divide $3x^4$ by $6x^6$.*

Arranging as above.

$$3x^4 \div 6x^6 = \frac{3x^4}{6x^6}$$

$$= \frac{3 \times x \times x \times x \times x \times x \times x}{6 \times x \times x \times x \times x \times x \times x \times x \times x \times x}.$$

In this example as the higher power is in the denominator, on cancelling there are $(6 - 4)$ factors in the denominator. Hence we get:

$$\frac{1}{2x^{6-4}} = \frac{1}{2x^2}.$$

Exercise 4.

Write down answers to the following:

1. $a^3 \div a$.
2. $a^4 \div a^2$.
3. $3x^3 \div x^2$.
4. $b^5 \div 2b$.
5. $6a^6 \div 3a^2$.
6. $5y^4 \div y$.
7. $6x^7 \div 2x^4$.
8. $14c^6 \div 7c^4$.
9. $a^2b^3 \div ab^2$.
10. $x^5y^4 \div x^2y^3$.
11. $5a^3b^3 \div ab^2$.
12. $6x^4y \div 2x^3$.
13. $6a^4 \div 3a^4$.
14. $15x^3 \div 12x^3$.

___

22. Easy fractions.

Algebraic fractions obey the same fundamental laws as fractions in Arithmetic. In principle they are manipulated by the same methods. But since the numerators and denominators may be algebraical expressions, sometimes rather complicated, they present difficulties not found in Arithmetic fractions. In this chapter we shall deal only with simple forms involving easy manipulation, more difficult cases being left until later.

23. Addition and subtraction.

The methods of Arithmetic can readily be applied, as shown in the following examples.

Example 1. *Find the sum of $\frac{x}{3} + \frac{x}{5}$.*

This is of the same form as

$$\frac{2}{3} + \frac{2}{5}$$

and is worked in the same way.

Just as

$$\frac{2}{3} + \frac{2}{5} = \frac{(2 \times 5) + (2 \times 3)}{3 \times 5}$$

so

$$\frac{x}{3} + \frac{x}{5} = \frac{(x \times 5) + (x \times 3)}{3 \times 5}$$

$$= \frac{5x + 3x}{15} = \frac{8x}{15}.$$

Example 2. *Find the sum of $\frac{3}{a} + \frac{4}{b}$.*

This is similar in type to the preceding and is dealt with in the same way:

$$\frac{3}{a} + \frac{4}{b} = \frac{(3 \times b) + (4 \times a)}{ab}$$

$$= \frac{3b + 4a}{ab}.$$

Example 3. *Simplify $\frac{x}{y} - \frac{a}{b}$.*

Proceeding as before

$$\frac{x}{y} - \frac{a}{b} = \frac{(x \times b) - (a \times y)}{y \times b}$$

$$= \frac{bx - ay}{by}.$$

Example 4. *Simplify* $\dfrac{2a}{15} + \dfrac{5b}{12}$.

As in Arithmetic, we find the L.C.M. of the denominators —viz. 60.

Then
$$\frac{2a}{15} + \frac{5b}{12} = \frac{(2a \times 4) + (5b \times 5)}{60}$$
$$= \frac{8a + 25b}{60}.$$

Note.—It is not possible to cancel any factors of 8 or 25 with factors of 60. This mistake is sometimes made by beginners, although the proceeding is contrary to the laws of Arithmetic. *Only factors common to each term of the numerator can be cancelled with factors of the denominator.*

Example 5. *Simplify* $\dfrac{x}{12a^2b} - \dfrac{y}{18ab^2}$.

We must find the L.C.M. of the denominators. To do this we find the L.C.M. of the numerical coefficients 12 and 18—*i.e.*, 36—then the L.C.M. of a^2b and ab^2. This is a^2b^2, since both of them will divide into it exactly. The product of 36 and a^2b^2 is the L.C.M. of the denominators.

$$\therefore \quad \frac{x}{12a^2b} - \frac{y}{18ab^2} = \frac{(x \times 3b) - (y \times 2a)}{36a^2b^2}$$
$$= \frac{3bx - 2ay}{36a^2b^2}.$$

24. Multiplication and division.

As in Arithmetic, these operations are based upon the same important rule of fractions, viz.:

If numerator and denominator are divided by the same number, the value of the fraction is unaltered.

This can be expressed algebraically as follows:

Let $\dfrac{a}{b}$ be any fraction.

Then
$$\frac{a}{b} = \frac{a \times n}{b \times n}$$

and
$$\frac{a}{b} = \frac{a \div m}{b \div m},$$

where m and n are any numbers.

Example 1. *Simplify* $\dfrac{4x^3y}{6xy^3}$.

Written in full the fraction is:

$$\frac{4x^3y}{6xy^3} = \frac{4 \times x \times x \times x \times y}{6 \times x \times y \times y \times y}.$$

Cancelling common factors this is equal to:

$$\frac{2 \times x \times x}{3 \times y \times y} = \frac{2x^2}{3y^2}.$$

Note.—In practice there is no need to write out the powers in full as shown above. The rule for the division of powers may be applied directly.

Example 2. *Simplify* $\dfrac{6ax^4}{14x^2y^2} \times \dfrac{2y^3}{3a^4}$

As in Arithmetic, factors in either numerator can be cancelled with factors in either denominator.

$$\therefore \qquad \frac{6ax^4}{14x^2y^2} \times \frac{2y^3}{3a^4}$$

$$= \frac{x^{4-2}}{7} \times \frac{2y^{3-2}}{a^{4-1}}$$

$$= \frac{x^2 \times 2y}{7 \times a^3}$$

$$= \frac{2x^2y}{7a^3}.$$

Example 3. *Simplify* $\dfrac{8x^3}{5a^2y} \div \dfrac{4x^2}{3a}$.

Proceeding as with a similar arithmetical example:

$$\frac{8x^3}{5a^2y} \div \frac{4x^2}{3a} = \frac{8x^3}{5a^2y} \times \frac{3a}{4x^2}$$

$$= \frac{2x \times 3}{5ay}$$

$$= \frac{6x}{5ay}.$$

Exercise 5.

Simplify the following:

1. $\dfrac{a}{5} + \dfrac{a}{7}.$

2. $\dfrac{2x}{3} - \dfrac{3x}{5}.$

3. $\dfrac{4}{x} + \dfrac{5}{y}.$

4. $\dfrac{3a}{2b} + \dfrac{5a}{3b}.$

5. $\dfrac{9x}{10y} - \dfrac{5x}{8y}.$

6. $\dfrac{7x}{3y} + \dfrac{2x}{9y} - \dfrac{5x}{6y}.$

7. $\dfrac{5a}{6bc^2} + \dfrac{3b}{8a^2c}.$

8. $x + \dfrac{4}{3x}.$

9. $3ab + \dfrac{5}{2b}.$

10. $8 - \dfrac{6}{5y^2}.$

11. $a + 1 + \dfrac{1}{a}.$

12. $x - 1 + \dfrac{1}{x}.$

13. $\dfrac{1}{ab} + \dfrac{2}{bc} + \dfrac{3}{ac}.$

14. $\dfrac{5}{a^2bc} - \dfrac{4}{ab^2c} + \dfrac{2}{abc^2}.$

---

15. $\dfrac{x^4y^2}{x^2y}.$

16. $\dfrac{12a^2b}{3a^4}.$

17. $\dfrac{15a^2b^4c}{10a^4b^2c^2}.$

18. $\dfrac{x^3 \times x^3}{x^6}.$

19. $\dfrac{4x^4 \times 2x^3y}{8x^6}.$

20. $\dfrac{a^4 \times bc^2}{a^3 \times b^2c^3}.$

21. $\dfrac{2a}{9} \times \dfrac{6}{a^2}.$

22. $\dfrac{x^2}{xy} \times \dfrac{3y^2}{2x}.$

23. $\dfrac{3p}{2q^2} \times \dfrac{6q}{5p}.$

24. $2x^2 \times \dfrac{y}{3x}.$

25. $2x \times \dfrac{x}{y}.$

26. $2x \div \dfrac{x}{y}.$

27. $\dfrac{x}{y} \div \dfrac{x}{y}.$

28. $\dfrac{x}{y} \div \dfrac{y}{x}.$

29. $\dfrac{3a}{7} \div \dfrac{9a}{14}.$

30. $\dfrac{2x^2}{y} \div \dfrac{3y}{x^2}.$

31. $\dfrac{6x^2}{8yz} \div \dfrac{3xy}{4xz}.$

32. $\dfrac{a^5}{b^4} \div \dfrac{3a^2}{2b^2}.$

33. $\dfrac{4abc}{5b^2} \div \dfrac{3c^2}{10ab}.$

34. $3xyz \div \dfrac{4xz}{y}.$

35. $30 \div \dfrac{3xyz}{2z}.$

36. $\dfrac{ab}{c^2} \div \dfrac{c^2}{4ab}.$

CHAPTER III

BRACKETS AND OPERATIONS WITH THEM

25. Removal of brackets.

SIMPLE examples of the use of brackets as a convenient way of grouping numbers have already been considered. In this chapter we will examine extensions of their use, operations with them and the simplification of algebraical expression, which contain brackets, by removing them.

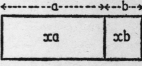

We will begin with a simple but important case.

Fig. 2 represents a rectangle made up of two other rectangles.

FIG. 2.

Let a mm = the length of one rectangle
  „ b mm =   „   „   the other rectangle.

Let x mm = the breadth of each rectangle as shown in Fig. 2.

Then $(a + b)$ mm = the length of the combined rectangle, placing $(a + b)$ in brackets shown in § 11.

The areas of two smaller rectangles are xa and xb mm² and area of whole rectangle = $x(a + b)$ mm².

But the area of the whole rectangle equals the sum of the area of the parts.

$$\therefore \qquad x(a + b) = xa + xb.$$

Similarly if there are three rectangles of lengths a, b and c ins. respectively, then in the same way we could show that:

$$x(a + b + c) = xa + xb + xc.$$

By modifications of the figures, which are left to the ingenuity of the student, we could similarly show that:

(1) $x(a - b) = xa - xb.$
(2) $x(a + b - c) = xa + xb - xc.$
(3) $x(a - b + c) = xa - xb + xc.$

39

In all these examples expressions containing brackets have been transformed into expressions without brackets, or, as we say, *the brackets have been removed*.

Hence we can deduce that:

When the whole of an expression within brackets is multiplied by any number, then, if the brackets be removed, each term within the brackets must be multiplied by the number.

The factor without the brackets is said to be *distributed* as a factor of each term within the brackets.

This is an example of the algebraical law called the **Law of distribution**.

26. Addition and subtraction of expressions within brackets.

There are four cases which we will consider. They are represented by the following expressions:

(1) $a + (b + c)$.
(2) $a + (b - c)$.
(3) $a - (b + c)$.
(4) $a - (b - c)$.

The question to be considered is, What will be the effect of removing the brackets in the above expressions, a, b and c representing any numbers?

(1) $a + (b + c)$

Let a, b, c be represented by the areas of rectangles as shown in Fig. 3.

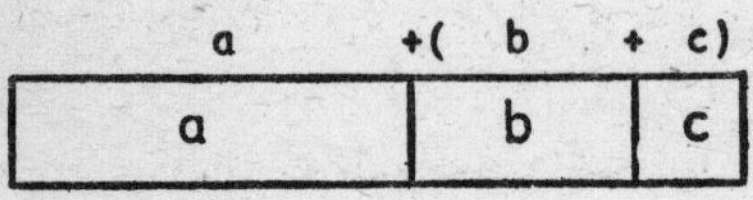

FIG. 3.

It is evident that the area of the whole rectangle which represents $a + (b + c)$ is the sum of the three rectangles representing a, b, c.

$$\therefore \qquad a + (b + c) = a + b + c.$$

The steps in the addition of the numbers in this case are not altered by the insertion or the removal of the brackets.

(2) $a + (b - c)$

As in the previous case, a, b, c are represented by the areas of rectangles as shown in Fig. 4.

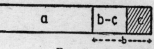

FIG. 4.

The two unshaded rectangles represent a and $b - c$.

∴ the whole unshaded portion represents $a + (b - c)$. It can be formed either:

    (1) by adding $(b - c)$ to a.
    (2) by adding b to a and then subtracting c.

These two results are equal.

∴ $a + (b - c) = a + b - c.$

Thus no change results when the brackets are removed.

FIG. 5.

(3) $a - (b + c)$

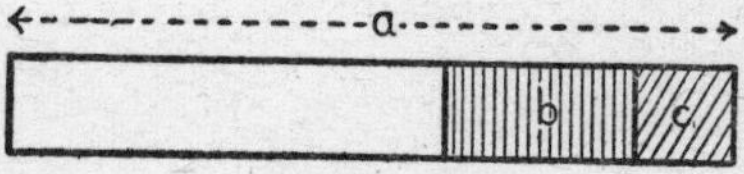

Using the same method as before of representing a, b and c, the unshaded rectangle (Fig. 5) represents the remainder when $(b + c)$, the two shaded rectangles, are subtracted from a, the whole rectangle, *i.e.*, it represents $a - (b + c)$.

Also if from a, the whole rectangle, we subtract b and c in turn the remaining rectangle is the unshaded portion.

∴ it represents $a - b - c$.

∴ $a - (b + c) = a - b - c.$

(4) $a - (b - c)$

In Fig. 6 the rectangle representing a, b, c is shown, a being represented by the whole rectangle.

The shaded rectangle represents $(b - c)$.

The unshaded rectangle represents the result of subtracting

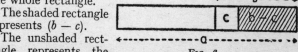

FIG. 6.

$(b - c)$ from a, *i.e.*, it represents $a - (b - c)$.

It may also be considered as representing the result of subtracting b from a and then *adding* c, *i.e.* it represents $a - b + c$.

$$\therefore \qquad a - (b - c) = a - b + c.$$

Collecting the four cases we have:

(1) $a + (b + c) = a + b + c.$
(2) $a + (b - c) = a + b - c.$
(3) $a - (b + c) = a - b - c.$
(4) $a - (b - c) = a - b + c.$

From these results we can deduce two rules respecting signs when the brackets are removed.

A. From (1) and (2) when the $+$ sign precedes the brackets the signs of the terms within the brackets are unaltered.

B. From (3) and (4) when the $-$ sign precedes the brackets the signs of the terms within the brackets are changed.

27. Worked examples.

The following are examples of the use made of the rules of § 25 and § 26 when brackets are removed from an algebraic expression in order to simplify it.

Example 1. *Simplify $a(a^2 + ab + b^2)$.*

When the brackets are removed the rule of § 25 is employed and the factor a multiplies each term within the brackets.

Thus: $\qquad a(a^2 + ab + b^2) = a^3 + a^2b + ab^2.$

Example 2. *Simplify $2(4a + 3b) + 6(2a - b)$.*

When removing the brackets we use § 25 to multiply by 2 and 6, and from § 26, since the $+$ sign before the second pair of brackets is positive, there is no change of sign.

$$2(4a + 3b) + 6(2a - b)$$
$$= 8a + 6b + 12a - 6b$$
$$= 20a.$$

Example 3. *Simplify $5x - (5y + 2x)$.*

This is an example of case (3) of § 26. On removing the brackets signs are changed.

Thus:
$$5x - (5y + 2x)$$
$$= 5x - 5y - 2x$$
$$= 3x - 5y \text{ (on adding like terms).}$$

Example 4. *Simplify* $3(4a - b) - 2(3a - 2b)$.

This involves the rule of § 25 and case (4) of § 26.
Using these
$$3(4a - b) - 2(3a - 2b)$$
$$= 12a - 3b - 6a + 4b$$
$$= 6a + b.$$

Example 5. *Simplify* $x(2x - y) - x(x - y) - y(x + 2y)$
and find its value when $x = 2, y = 1$.

$$x(2x - y) - x(x - y) - y(x + 2y)$$
$$= 2x^2 - xy - x^2 + xy - xy - 2y^2$$
$$= x^2 - xy - 2y^2 \text{ (since } + xy - xy = 0).$$

Substituting $x = 2, y = 1$.

$$x^2 - xy - 2y^2 = (2)^2 - (2 \times 1) - 2(1)^2$$
$$= 4 - 2 - 2$$
$$= 0.$$

Note.—This example shows the advantage of simplifying the expression before substituting the values of x and y.

Exercise 6.

Simplify the following expressions by removing brackets.

1. $3(5x + 6z)$.
2. $2a(3a + 4b)$.
3. $6a^2(3a + 7b - 6c)$.
4. $2(x + 2y) + 3(2x - y)$.
5. $x(x^2 - 3x) + x^2(4x + 7)$.
6. $\frac{1}{2}(x - 2y) + \frac{1}{4}(2x + 4y)$.
7. $2(x + y + z) + 3(2x + y - 2z)$.
8. $x - (2y + z)$.
9. $2x - (y - 2z)$.
10. $2(2a + 2b) - 3(a - b)$.
11. $3a - (2a + b)$.
12. $3a - (2a - b)$.
13. $5x - (x - 2y + 2z)$.
14. $3(a + b - c) - 2(a - b + c)$.
15. $4(x + y) - 3(2x - y) + 2(x - 2y)$.

16. $a(a + b) - b(a - b)$.
17. $x^2(x + y) - xy(x^2 - y^2)$.
18. $3(x^2 + x + 5) - 2(x^2 - 3x - 4)$.
19. $2p(3p + 2q) - 3q(2p - 5q) + p(3p + 5q)$.
20. $5(xy)^2 - 3x(y - 2x)$.
21. $(2x^2)^2 - 2x^3(x - 4)$.

28. Systems of brackets.

It may happen that an expression within brackets is part of another expression which is itself within brackets. In that case a second set of brackets would be required, and to avoid confusion they must be of a shape different from those already used, such as $\{$——$\}$ or $[$——$]$.

For example:

$$40 - \{2(a + b) + 5(a - b)\}.$$

The student will easily recognise how clearly and effectively the brackets help to show the construction of the expression and relations of the different parts to one another.

It might happen that the whole of the above expression is to be multiplied by $2b$. This will necessitate another set of brackets which will indicate that the expression is to be treated as a whole. We would express this as follows:

$$2b[40 - \{2(a + b) + 5(a - b)\}].$$

Sometimes a straight line may be placed over a part of an expression with the same meaning as brackets.

Thus $x - \overline{y + z}$ has the same meaning as $x - (y + z)$. It must also be remembered that an expression which is the numerator of a fraction must be regarded as a whole and must be treated as if it were in a bracket.

Thus: $$5a - \frac{a - b}{2}$$

means the same as $\quad 5a - \frac{1}{2}(a - b)$.

If the whole expression were multiplied by 2, it would become $10a - (a - b)$.

When expressions with two or more sets of brackets are to be simplified by removal of the brackets, it is well, as a rule, to begin with the innermost and work outwards. Examples of this will be seen in the following.

29. Worked examples.

Example 1. *Simplify* $2\{3a + 5(b + c)\}$.

As stated above we begin by removing the inner brackets.

$$2\{3a + 5(b + c)\}$$
$$= 2\{3a + 5b + 5c\}$$
$$= 6a + 10b + 10c.$$

Example 2. *Simplify* $3\{3a - 2(a - b)\}$.

$$3\{3a - 2(a - b)\}$$
$$= 3\{3a - 2a + 2b\}$$
$$= 3\{a + 2b\}$$
$$= 3a + 6b.$$

Example 3. *Simplify* $12a - 2[3a - \{4 - 2(a - 3)\}]$.

Beginning with the innermost bracket

$$12a - 2[3a - \{4 - 2(a - 3)\}]$$
$$= 12a - 2[3a - \{4 - 2a + 6\}]$$
$$= 12a - 2[3a - \{10 - 2a\}] \quad \text{(adding like terms)}$$
$$= 12a - 2[3a - 10 + 2a]$$
$$= 12a - 2[5a - 10]$$
$$= 12a - 10a + 20$$
$$= 2a + 20.$$

Exercise 7.

Remove the brackets from the following expressions and simplify them.

1. $3\{5a - 3(a + 1)\}$.
2. $3\{4(a + b) - 3(a - 2b)\}$.
3. $\frac{1}{2}\{6x - 3(2x - 1)\}$.
4. $5a^2 + 2a\{b - (a + c)\}$.
5. $3p^2 - \{2p^2 - p(p + 1)\}$.
6. $3x(x + 3y) - 2\{x^2 + 3y(x - 2y)\}$.
7. $3bc - 2\{b(b - c) - c(b + c)\}$.
8. $15x - [3x - \{2x - (x - 5)\}]$.
9. $50 - 2[3a + 2\{3b - 4(b - 1)\}]$.
10. $2(x + y) - x - y$.
11. $12\left\{\dfrac{a - b}{2} - \dfrac{a + b}{3}\right\}$.

12. $3x\left\{\dfrac{2x-y}{3}-\dfrac{x-2y}{6}\right\}.$

13. $3c-\left\{\dfrac{2a+c}{8}-\dfrac{a+2c}{4}\right\}.$

Fill in the blanks within the brackets in the following:

14. $2a-b+c=2a-($ $).$
15. $x-y-z=x-($ $).$
16. $2a+4b-6c=2($ $).$
17. $x^2-xy+y^2=x^2-y($ $).$

---

18. From $3a-2b+4c$ subtract $a+2b-3c$.
19. Take $2x-3y+4z$ from $3x-y+2z$.
20. When $a=3$, $b=2$, $c=1$, find the values of the following:

    (1) $4a(a+4b)-a(3a-b).$
    (2) $3c\{4c-(3c-1)\}.$

CHAPTER IV

POSITIVE AND NEGATIVE NUMBERS

30. The scale of a thermometer.

FIGS. 7a and 7b represent portions of Celsius thermometers in which a fine column of mercury registers the rise and fall of temperature.

The zero point, marked 0, indicates the position of the mercury in the tube at freezing point—*i.e.*, the freezing point of water.

Fig. 7(a) shows the mercury at 8° C **above zero.**

Now suppose the temperature falls 16 C° below this point.

First it falls 8 C° to 0° C, and then continues to fall for 8 C° **below zero.** To show this temperature on the scale it must be marked in some way which is different from the 8° C **above zero,** or there would be confusion. To distinguish the degrees below zero from those above we use the plan of putting a minus sign, —, before all those below zero, and if necessary a plus sign, +, before those above zero.

Thus + 8° C means 8 degrees **above zero,** and − 8° C ,, ,, ,, **below zero.**

We may call these **positive** and **negative** degrees, using the terms, and the signs + and —, to indicate **different directions** up and down from the zero.

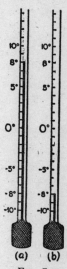

(a) (b)

FIG. 7.

31. An example from time.

In reckoning time, the numbers denoting the years are counted from the birth of Christ. Years after that event are denoted by A.D., as A.D. 1941, and those before by B.C., as 55 B.C. The use of the symbols A.D. and B.C. is, in principle, similar to the use of + and − in the case of the thermometer.

32. A commercial illustration.

In a certain transaction at market a farmer made a profit of £12.

In a second deal he lost £8.

Consequently in the two transactions he made a net gain of £12 − £8 = £4.

In a third transaction he lost £10.

His profit and loss account is now shown by £4 − £10.

If this loss had been £4, his position would be £4 − £4 = 0—*i.e.*, he has reached a zero position, neither loss nor gain.

But he lost £10, not £4; therefore he has a net loss of £6—*i.e.*, he is £6 below his zero.

To distinguish gains from losses, we could, as in the case of the thermometer, place the

negative sign before amounts showing losses,

and the

positive sign before amounts showing profits.

In that sense £4 − £10 = − £6, the negative sign indicating a loss of £6.

33. Motion in opposite directions.

Suppose a man starting from a point O (*see* Fig. 8) travels for 4 km in the direction O to X, reaching the point marked A.

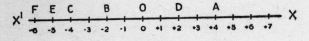

Fig. 8.

He then turns and travels 6 km in the opposite direction, O to X¹. After 4 km he reaches O, his **zero** or starting point. The next 2 km take him to B.

He is now 2 km from O but in the direction **opposite** to that in which he started. His successive distances from O can be shown by + 4 − 6. This suggests that, as in the previous cases, if distances from O in one direction were regarded as positive, the distances in the opposite direction could be regarded as negative.

Thus, if we now say that the man is − 2 km from O, we mean that he is 2 km in the direction opposite to the original. Accordingly, in the diagram showing the movements from O (Fig. 8)

Distances to the right with + signs we would call positive.

,,   ,,   left with − signs   ,,   ,, negative.

With this device, when giving his position from O, the sign of the number would indicate in which direction the man is from O.

Thus − 4 km would indicate he is at C, + 2 km would show he is at D.

The number with the + sign we call a positive number.

,,   ,,   − sign   ,, negative number.

34. Positive and negative numbers.

From this it appears that we have devised a new kind of number—viz., a negative number—and that, in consequence, we can divide numbers into two kinds: positive and negative. From the examples above a negative number is a number which in its meaning and effect is opposite to a positive number.

Frequently, as in examples of §§ 30 and 33, the negative number indicates a direction opposite to that of the positive number, and in this sense,

positive and negative numbers are called directed numbers.

If negative numbers can rightly be classed as numbers, they must, in operations with them, conform to the rules governing the numbers which we now call positive numbers. These operations will be considered fully later, but a few simple illustrations will serve to show that we can deal with them in the same way as positive numbers.

For example, in the matter of addition, we can add − 2 and − 3, and a glance at Fig. 8 will show that the result is − 5, being equal to the sum of − 2 and another − 3 from O to the point marked E.

Or if − 3 be multiplied by 2—*i.e.*, we double the distance from O—we get − 6, at the point F.

Similarly division of -6 by 2 would give -3.

For the rest of this chapter, in order to make the meaning clear, positive and negative numbers, when being used in operations will be placed in brackets.

Thus $(-6) \div (+2) = (-3)$.

35. Negative numbers.

Corresponding to every positive number there is a negative number, and we can write a succession of negative numbers corresponding to positive numbers.

Thus if we write down the numbers beginning, for example, with $+6$ and decreasing by one at each step, we get the series of numbers $+6, +5, +4, +3, +2, +1, 0$.

With the negative number, we do not stop at the zero, but continue with the subtraction, so that we get

$$-1, -2, -3, -4, -5 \ldots, \text{etc.}$$

In descending order of magnitude. Or if we start with (-6) and add unity in succession we get the complete series :

$$-6, -5, -4, -3, -2, -1, 0\ +1, +2, +3, +4, +5 \ldots,$$

in ascending order of magnitude. This series can be extended in either direction and decimals and fractions fall into their places between these numbers. Thus we get what is called the complete number scale.

36. Graphic representation of the complete number scale.

The graphical representation of the complete number scale is so important that we return to it again, using square ruled paper.

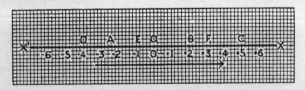

Fig. 9.

The straight line XOX' is drawn, as in Fig. 9, to show the representation of a small part of the scale.

On this line, starting from a point O, and using a suitable scale, distances are marked to the right to represent positive numbers and to the left to represent negative numbers.

We could imagine this line to be extended to any distance on either side so that any number could be included. Numbers involving decimals lie between those marked. Thus $- 2·5$ would be at A.

Two principles may be noted:

(1) *Every number can be represented at its appropriate point on the scale.*

(2) *Conversely, every point on the scale represents a number.*

It should be observed that the numbers represented in the figure increase from left to right, as shown by the arrow.

37. Operations with negative numbers.

With the introduction of negative numbers, Algebra passes beyond the boundaries of Arithmetic. We must therefore proceed to examine operations with this new kind of number, remembering, as previously stated, that it must conform to the laws of Algebra. The important operations for our present consideration are the fundamental ones of Addition, Subtraction, Multiplication and Division.

38. Addition of positive and negative numbers.

We have already seen in § 34 that addition of two negative numbers is performed in the same way as that of positive numbers.

Just as
$$(+ 3) + (+ 2) = + 5$$
so
$$(- 3) + (- 2) = - 5.$$

Such operations can be confirmed by use of Fig. 9.

The addition of a positive and a negative number can also be seen from Fig. 9.

For example, $(- 4) + (+ 3)$ is represented by starting at D, which represents $- 4$, and moving $+ 3$ to the right to E, the result being $- 1$.

Similarly $(+ 3) + (- 7)$ as found by starting at F,

marking $+ 3$, and since $- 7$ is a negative number, we move 7 divisions to the left to D to find the sum, which is $(- 4)$.

When the negative numbers involve letters, the procedure is the same. Thus:

$$(- 5a) + (+ a) = - 4a$$
$$(+ 2b) + (- 5b) = - 3b$$
$$(- 2x) + (+ 6x) = 4x.$$

39. Subtraction.

This operation presents a little more difficulty, since it is not easy at first to understand what is implied by the subtraction of a negative number, as, for example, $(+ 6) - (- 2)$ or $(- 2) - (- 5)$.

This can be deduced from Fig. 9, but we will first obtain the rule by applying a fundamental law of addition and subtraction.

Since $\qquad 9 = 7 + 2$
then $\qquad 9 - 2 = 7$
and $\qquad 9 - 7 = 2,$

or, in general terms: if

$$a = b + c$$
then $\qquad a - b = c$
and $\qquad a - c = b.$

We have seen above that

$$(- 5) + (+ 3) = - 2$$
or $\qquad (- 2) = (- 5) + (+ 3).$

$\therefore$ from the above

$$(- 2) - (- 5) = + 3$$
but we know that $(- 2) + (+ 5) = + 3.$

$\therefore$ comparing the two statements

$$- (- 5) = (+ 5).$$

A similar result will clearly hold whatever the numbers.
$\therefore$ We conclude that for any number a

$$- (-a) = + a.$$
Similarly $\qquad - (- 2a) = + 2a.$

Examples.

$$5x - (-3x) = 5x + 3x = 8x$$
$$-2b - (-4b) = -2b + 4b = 2b.$$

Graphical illustrations.

To find $(-2) - (-5)$.

The rule can be deduced from the graphical representation of the number scale in Fig. 9 as follows:

If we **add** a negative number we move to the **left** along the scale.

Thus $\qquad (-2) + (-5) = -7.$

Consequently if we **subtract** a negative number we must move to the **right**.

Starting from (-2) and moving 5 to the right we reach $+3$, *i.e.*, $\qquad (-2) - (-5) = (+3)$.

To find $(-2) - (+5)$.

When adding a positive number we move to the **right**.

∴ when **subtracting** a positive number we move to the **left**.

∴ starting from (-2) we move 5 divisions to the left and read (-7).

Summarising the rules for addition and subtraction we have:
$$+ (+a) = (+a)$$
$$+ (-a) = (-a)$$
$$- (+a) = (-a)$$
$$- (-a) = (+a).$$

The student should compare these with rules for signs given in § 26.

Exercise 8.

1. A lift starting from the ground-floor rises to the fourth floor. Then it descends to the second floor, rises to the sixth floor and finally descends to the ground-floor. Express its movements by using positive and negative numbers.

2. The movement of the mercury in a thermometer was as follows. Starting at $+8°$ C it rose 2 C°, fell 14 C°, then rose 4 C° and finally fell 6 C°. Express these, using positive and negative signs, and find the final temperature.

3. How much higher than a temperature of $-15°$ C is:

    (1) A temperature of $-4°$ C?
    (2) Freezing point?
   •(3) $+15°$ C?

4. Using the number scale shown in Fig. 9 find:

    (1) By how much -2 is greater than -7?
    (2) By how much -6 is less than -1?
    (3) By how much $+3$ is greater than -5?

5. (1) What must be added to (-3) to give (a) -1, (b) $+1$?

    (2) What must be taken from (-3) to give (-8)?

6. Write down the values of:

    (1) $(+6) + (-2)$.　　　　(2) $(+6) - (-2)$.
    (3) $(-6) + (-2)$.　　　　(4) $(-6) - (-2)$.
    (5) $0 - (-3)$.　　　　　(6) $(-4) + (-4)$.
    (7) $-(4) - (-4)$.　　　　(8) $(-4) - (+4)$.

7. Simplify the following:

    (1) $+2a - (-5a)$.
    (2) $-4x - (+3x)$.
    (3) $+3ab - (-7ab)$.
    (4) $2x - 3y - 5y - 3x$.
    (5) $(3a - 2b) - (2a + 5b)$.
    (6) $(3x - y) - (4x - 3y)$.
    (7) $3x - (3y - 4x)$.
    (8) $(5 + x) - (6 - 2x) - (3x + 7)$.

8. (2) Subtract $(x - 2y)$ from $(3x - 4y)$.
   (2) Subtract $(x - y + 2z)$ from $(3x - 2y - 5z)$.

9. Fill in the brackets in the following:

    (1) $3a - (\quad) = 8a$.
    (2) $5x - (\quad) = -x$.
    (3) $-3a - (\quad) = 7a$.

10. Write down the values of:

    (1) $0 - (+a)$.　　　　(2) $0 - (-a)$.
    (3) $0 + (-a)$.

40. Multiplication.

(1) Multiplication of $(-a)$ by $(+b)$
and $\quad\quad\quad\quad\quad\quad (+a)$ by $(-b)$.

Consider, as a special case $(-2) \times (+3)$.

Since multiplication is a shortened form of addition, the meaning of $(-2) \times (+3)$ is $(-2) + (-2) + (-2)$.

This by § 59 $= (-6)$.

∴ $\quad\quad\quad\quad (-2) \times (+3) = (-6)$.

This can be applied to any pair of numbers, and so we may conclude that in general

$$(-a) \times (+b) = (-ab).$$

Since the multiplication of two numbers can be taken in any order (§ 17)

$$(+a) \times (-b) = (-b) \times (+a)$$

and this by the above result is $(-ab)$.

(2) Multiplication of $(-a)$ by $(-b)$.

Since $\quad\quad\quad (-a) \times (+b) = (-ab)$

and a negative number operates in the opposite sense to a positive number, it follows that

$$(-a) \times (-b) = (+ab).$$

41. Division.

(1) Division of $(+a)$ by $(+b)$.

Since $\quad\quad\quad (+4) \times (+3) = +12$.

∴ $\quad\quad\quad (+12) \div (+4) = (+3)$.

Similarly $\quad + (a) \div (+b) = +\left(\dfrac{a}{b}\right)$.

(2) Division of $(-a)$ by $(+b)$.

Since $\quad\quad\quad (+4) \times (-3) = (-12)$.

∴ $\quad\quad\quad (-12) \div (+4) = (-3)$.

Similarly $\quad (-a) \div (+b) = \left(-\dfrac{a}{b}\right)$.

(3) Division of $(+a)$ by $(-b)$.

Again $\quad\quad\quad (-4) \times (-3) = (+12)$
and $\quad\quad\quad (+12) \div (-4) = (-3)$.

Similarly $\quad (+a) \div (-b) = \left(-\dfrac{a}{b}\right)$

(4) **Division of** $(-a)$ **by** $(-b)$.

As above $(+4) \times (-3) = (-12)$.

$\therefore$ $(-12) \div (-3) = (+4)$,

and in general $(-a) \div (-b) = \left(+\dfrac{a}{b}\right)$.

42. Summary of rules of signs for multiplication and division.

Multiplication.

$(+a) \times (+b) = +ab$
$(+a) \times (-b) = -ab$
$(-a) \times (+b) = -ab$
$(-a) \times (-b) = +ab$.

Division.

$$(+a) \div (+b) = \left(+\frac{a}{b}\right)$$

$$(+a) \div (-b) = \left(-\frac{a}{b}\right)$$

$$(-a) \div (+b) = \left(-\frac{a}{b}\right)$$

$$(-a) \div (-b) = \left(+\frac{a}{b}\right)$$

These results can be summarised in the following rule:

In the multiplication and division of positive and negative numbers, if the two numbers have the same sign the result is a positive number. If the signs are different, the result is a negative number.

Or to remember these more readily the following slogan can be used:

Like signs give +
Unlike signs give −

43. Powers. Squares and square roots.

When we square a number we multiply two numbers with the same signs. In accordance with the above rules, the product must be positive.

Thus
$$(+ a) \times (+ a) = + a^2$$
$$(- a) \times (- a) = + a^2.$$

Consequently the **square of any number is positive**.

It follows that when this operation is reversed and the square root of a^2 is required, this may be either $(+ a)$ or $(- a)$.

To indicate this we use the sign $\pm$, meaning " plus or minus ", *i.e.*,

$$\sqrt{a^2} = \pm a.$$

Again

$$(- a)^3 = (- a) \times (- a) \times (- a) = - a^3$$

and

$$(- a)^4 = (- a) \times (- a) \times (- a) \times (- a) = + a^4.$$

From these and similar examples we may deduce that:

An *odd power of a negative number is negative*.
An *even power of a negative number is positive*.

Exercise 9.

1. Write down the answers to the following:

    (1) $(+ 12) \times (+ 3)$.   (2) $(+ 12) \times (- 3)$.
    (3) $(- 12) \times (+ 3)$.   (4) $(- 12) \times (- 3)$.
    (5) $(+ 12) \div (+ 3)$.   (6) $(+ 12) \div (- 3)$.
    (7) $(- 12) \div (+ 3)$.   (8) $(- 12) \div (- 3)$.

2. Write down the answers to the following:

    (1) $(+ a) \times (- a)$.   (2) $(- a) \times (- a)$.
    (3) $(+ a) \div (- a)$.   (4) $(- a) \div (- a)$.

3. Write down the answers to the following:

    (1) $(- 2a) \times (+ 2b)$.   (2) $(- 2a) \times (- 2b)$.
    (3) $(+ 10x) \times (- 2y)$.   (4) $(- 10x) \times (+ 2y)$.
    (5) $(+ 10x) \div (- 2y)$.   (6) $(- 10x) \div (- 2y)$.

4. Find the values of the following:

    (1) $(- 4) \times (+ 3) \times (- 2)$.
    (2) $(- a) \times (+ 3a) \times (- 2a)$.
    (3) $(- 18xy) \div (- 6x)$.
    (4) $(- 24a^2b^2) \div (+ 4ab)$.

5. Find the values of the following:

    (1) $(-5x) \times (-2x) \times (-x)$.
    (2) $a(a - b) - b(b - a)$.
    (3) $-\{a(-2b) \times (-b)\}$.
    (4) $-a(a - 2b - c)$.

6. Find the simplified form of the following:

    (1) $(+2a) \times (-5b) \times (-2b)$.
    (2) $(-4x)^2 - 2x(5x - 4)$.
    (3) $x(y - z) - z(x - y) - y(x - z)$.

7. Write down the second, third, fourth and fifth powers of:

    (1) $(-a)$.   (2) $(-2x)$.   (3) $\left(-\dfrac{b}{3}\right)$.

8. Write down the square roots of 81 and $9x^4$ and the cube roots of $-x^3$ and $-8a^6$.

9. Find the answers to the following:

    (1) $(-8x) \times (-2)$.
    (2) $(-10x) \div (-2)$.
    (3) $(-2xy) \div (-x)$.
    (4) $(+6b) \div (-3)$.
    (5) $(+8t^2) \div (-4t)$.
    (6) $(-4x^3) \div (-2x^2)$.
    (7) $(-2x^2) \times (-4x)$.
    (8) $(+15x^2y) \div (-5xy)$.
    (9) $(-12a^2b^2) \div (+3ab)$.
  (10) $(-24a^3bc^2) \div (-4abc)$.

10. Write down the values of:

    (1) $\{(-a) \times (-b)\} \div (-a)$.
    (2) $\left(+\dfrac{1}{x^2}\right) \times (-x)^3$.
    (3) $(-6x)^2 \times (-x)^3 \div (-2x)^4$.

CHAPTER V

SIMPLE EQUATIONS

44. Meaning of an equation.

If it is known that 5 times a certain number is 40, a simple process in Arithemtic enables us to calculate that the number is 8.

In algebraic form the problem could be expressed as follows:

Let n = the unknown number. Then the question can be put in this way:

If $5n = 40$, what is the value of n?

The statement $5n = 40$ is called an **equation**. It is a statement of equality, but it also implies that a value of n is required which will make the left-hand side of the equation equal to the right, or as we say "*satisfies the equation*". The process of finding the value of n which thus satisfies the equation is called "*solving the equation*".

The solution of the above equation involves no more than the division of the right-hand side by the coefficient of n, and could be stated thus:

$$5n = 40.$$
$$\therefore \quad n = 40 \div 5.$$
$$\therefore \quad n = 8.$$

The solution of an equation is rarely so simple as this. Equations usually consist of more or less complicated expressions on both sides of the equation. By various operations we aim at reducing the equation to the simple form above. The value of the unknown letter is then easily found. These operations will be illustrated in the examples which follow.

45. Solving an equation.

Example I. *If 8 times a number is decreased by 5 the result is 123. What is the number?*

59

This simple problem could be solved mentally, but it will serve as an introduction to the process of solution.

Let n = the number.

Then $8n - 5$ is the expression which states algebraically " 8 times the number decreased by 5 ".

But this is equal to 123.

Hence we can form the equation

$$8n - 5 = 123.$$

This is the first step that must always be taken—**to formulate the equation.** Then we proceed to the solution—*i.e.,* to find the value of n which satisfies it.

Now, the above statement means that 123 is 5 less than 8 times the number, or, if 123 be increased by 5, it is equal to 8 times the number.

∴ we can write the equation in this form

$$8n = 123 + 5.$$

Thus we have practically reached the form we wanted, after which we can readily find the solution.

This step was reached in effect, by transferring the 5 to the right-hand side, leaving only a multiple of n, the unknown number, on the left side. In this transference the argument involved **changing the sign of the 5.**

The same result could be obtained as follows:

Since $\qquad\qquad 8n - 5 = 123,$

if each side be increased by 5 the result will be that we shall be left with $8n$ only on the left-hand side and the two sides will still be equal. We shall have as our equation:

$$8n - 5 + 5 = 123 + 5$$

whence $\qquad\qquad 8n = 128$

and $\qquad\qquad n = 16.$

This device is employed in the solution of practically every equation and it depends on the fact that:

(A) **If the same number be added to, or subtracted from, both sides of an equation, the two sides will again be equal.**

As a working rule this is equivalent *to transferring a number from one side of an equation to the other at the same time changing its sign—i.e., change + to − and − to +.*

A principle similar to the above which will be employed later, is:

(B) If both sides of an equation are multiplied or divided by the same number, the two sides of the new equation will be equal. If the multiplier is — 1, both sides change signs.

Example 2. In § 4 we saw that three consecutive odd numbers could be expressed algebraically by

$$2n + 1, 2n + 3, 2n + 5$$

where n is any integer.

Now suppose we wish to solve this problem:

The sum of three consecutive odd numbers is 81. What are the numbers?

As stated above, the first step is to form an equation. This usually means putting into algebraic form the facts which are given about the unknown number or numbers.

We first, as above, represent the three odd numbers by

$$2n + 1, 2n + 3, 2n + 5.$$

Then we express algebraically, the fact that their sum is 81. Hence we get the equation:

$$(2n + 1) + (2n + 3) + (2n + 5) = 81.$$

The use of the brackets helps to make the statement clear.

We now remove the brackets and get:

$$2n + 1 + 2n + 3 + 2n + 5 = 81.$$

Adding like terms $\qquad\qquad\quad 6n + 9 = 81$

whence, as above $\qquad\qquad\quad\quad 6n = 81 - 9$

or $\qquad\qquad\qquad\qquad\qquad\quad 6n = 72.$

$\therefore \qquad\qquad\qquad\qquad\qquad\quad n = 12.$

We now obtain the odd numbers by substitution of $n = 12$ in $\qquad\qquad\quad 2n + 1, 2n + 3, 2n + 5.$

$\therefore$ the numbers are   **25, 27, 29.**

This should be checked by ascertaining that their sum is 81.

46. Worked examples.

Equations arise out of practical problems in a variety of ways, and examples will be given later, but it is desirable that the student should first have sufficient practice in the methods of solving equations. Examples of equations will therefore be worked out and provided for practice which will have no relation to any special problems.

It is usual in such practice equations to use letters at the end of the alphabet, x, y and z, to represent the unknown numbers, and, when necessary, letters at the beginning of the alphabet, a, b, c, etc., to represent known numbers.

This choice of letters is due to Descartes (seventeenth century).

Example 1. *Solve the equation:*

$$6x - 5 = 2x + 9.$$

The general plan adopted is to *collect the terms involving the unknown number, x, on the left side, and the other terms on the right*.

Transferring the x term from the right side we get:

$$6x - 2x - 5 = 9.$$

Transferring the -5,

$$6x - 2x = 9 + 5$$
$$\therefore \qquad 4x = 14$$
and
$$x = \frac{7}{2}.$$

Note.—With practice the two transference steps could be taken together.

Check. The accuracy of the solution to an equation can always be checked by substituting the value found in both sides of the original equation. In the above case:

Left side $\qquad \left(6 \times \dfrac{7}{2}\right) - 5 = 16.$

Right side $\qquad \left(2 \times \dfrac{7}{2}\right) + 9 = 16.$

The two sides are equal.

$$\therefore \qquad x = \frac{7}{2} \text{ satisfies the equation.}$$

Example 2. *Solve the equation:*
$$10(x - 4) = 4(2x - 1) + 5.$$

First simplify both sides by removing brackets.

Then $\qquad 10x - 40 = 8x - 4 + 5.$

Transferring $8x$ to the left side and $- 40$ to the right,

$$10x - 8x = 40 - 4 + 5.$$
$$\therefore \qquad 2x = 41$$
and $\qquad x = 20\tfrac{1}{2}.$

Check.

Left side $\qquad 10(20\tfrac{1}{2} - 4) = 10 \times 16\tfrac{1}{2} = 165.$

Right side $4(2 \times 20\tfrac{1}{2} - 1) + 5 = 160 + 5 = 165.$

$\therefore \qquad x = 20\tfrac{1}{2}$ satisfies the equation.

Example 3. *Solve the equation:*

$$\frac{3x}{5} + \frac{x}{2} = \frac{5x}{4} - 3.$$

When the equation involves fractions, the first step, in general, towards simplification is to " clear the fractions ". This is effected by multiplying throughout by such a number that the fractions disappear. This is justified by Principle B, § 45.

The smallest number which will thus clear the fractions is the L.C.M. of their denominators, in this case 20.

Multiplying *every* term on both sides by 20 we get:

$$\left(\frac{3x}{5} \times 20\right) + \left(\frac{x}{2} \times 20\right) = \left(\frac{5x}{4} \times 20\right) - (3 \times 20).$$

$$\therefore \qquad 12x + 10x = 25x - 60.$$
$$\therefore \qquad 12x + 10x - 25x = - 60$$
$$- 3x = - 60.$$
$$\therefore \qquad x = - 60 \div - 3.$$
$$\therefore \qquad x = 20 \quad \text{(by the law of signs).}$$

This should be checked as in previous examples.

Example 4. *Solve the equation:*

$$4x - \frac{x - 2}{3} = 5 + \frac{2x + 1}{4}.$$

Multiplying throughout by 12

$$48x - 4(x - 2) = 60 + 3(2x + 1). \quad (\S 28.)$$

Clearing brackets

$$48x - 4x + 8 = 60 + 6x + 3.$$
$$\therefore \quad 48x - 4x - 6x = 60 + 3 - 8.$$
$$\therefore \quad\quad\quad\quad 38x = 55$$

and
$$x = \frac{55}{38}.$$

Check by substitution.

Exercise 10.

Solve the following equations:

1. (a) $14x = 35$.   (b) $1{\cdot}5x = 30$.
2. (a) $\frac{3}{2}x = 24$.   (b) $\frac{5}{8}x = 50$.
3. (a) $3x = -48$.   (b) $-2x = 20$.
4. (a) $-\frac{1}{2}x = -40$.   (b) $-\frac{1}{3}x = 24$.

5. (a) $\dfrac{x}{0{\cdot}6} = 21$.   (b) $\dfrac{2x}{0{\cdot}2} = -0{\cdot}8$.

6. (a) $5x + 8 = 24$.   (b) $10x - 9 = 41$.
7. (a) $3x - 4{\cdot}7 = 2{\cdot}8$.   (b) $2{\cdot}5x + 50 = 80$.
8. (a) $3x - 5 = 2x + 3$.   (b) $6y + 11 = 3y + 15$.
9. (a) $2x + 6 = 14 - 3x$.   (b) $z + 20 = 5z - 44$.
10. (a) $3y - 1{\cdot}5 = 7y - 8{\cdot}7$.
   (b) $4{\cdot}8x + 52 = 3{\cdot}2x - 20$.

11. (a) $3y - \dfrac{44}{5} = 2y + 6$.   (b) $x + \dfrac{x}{15} = 20$.

12. (a) $\dfrac{3x}{5} - \dfrac{2}{3} = 0$.   (b) $\dfrac{x}{0{\cdot}1} + \dfrac{x}{0{\cdot}2} = 5$.

13. $4(2x - 5) = 3(2x + 8)$.
14. $3x - 2(x + 4) = 5x - 28$.
15. $2(x + 5) - 3(x - 6) = 20$.
16. $5(y - 1) - 2(y + 6) = 2y + 12$.
17. $2(x - 1) + 3(x + 4) = 4(x + 1) - (x - 5)$.
18. $3(x - 7) - 3(2x - 4) = 4(x + 3)$.
19. $\frac{1}{2}(3y + 6) - \frac{1}{3}(2y - 4) = 20$.

20. (a) $\dfrac{x}{8} - \dfrac{4}{5} = 0$.   (b) $\dfrac{2x}{5} - 3 = 8$.

21. (a) $\dfrac{7x}{8} - (x - 2) = 12$.   (b) $\dfrac{3x}{2} - \dfrac{x}{3} = \dfrac{5(x - 4)}{6}$.

22. $\dfrac{2x - 5}{3} - \dfrac{3x - 1}{4} = \dfrac{3}{2}$.

23. $\dfrac{7x-6}{4} - \dfrac{5x+3}{7} - \dfrac{6x-1}{14} = 3.$

24. $\frac{1}{2}(x+3) - \frac{2}{5}(x-3) = x + 9.$

25. $12(5-x) - 3(3x-4) = 23.$

26. $\frac{3}{2}(x-1) - \frac{1}{3}(5-x) = 2\frac{1}{3} - 3(x-3).$

27. $\dfrac{y}{7} - \dfrac{1-y}{5} = 2 - \dfrac{3y}{5}.$

28. $\dfrac{2x+4}{1\cdot5} - \dfrac{3x-5}{2\cdot5} = 1.$

29. Solve for n the equation $2n = 0\cdot58(12 - n)$.

30. For what value of r is $18\cdot4$ equal to $2(3\cdot5r - 1)$?

31. Find x when $\dfrac{7\cdot5}{x} = \dfrac{5}{2}.$

32. Find c if $\dfrac{18}{2c} = 3\cdot8.$

33. If $C = \dfrac{V}{R}$, find V when $C = 8$, $R = 4\cdot5$.

34. For what value of x is $3(x-5)$ equal to $\dfrac{4x+3}{2}$?

---

47. Problems leading to simple equations.

The methods of solving problems by means of simple equations are illustrated by the following examples. The general method of procedure is:

(1) Having decided which is the unknown quantity, represent it by a letter, such as x, stating clearly the units employed when necessary.

(2) Form an equation which represents the facts provided by the problem about the unknown quantity.

(3) Solve the equation.

Example 1. *At a dance party there were 10 more women than men. The men paid 20p each, the women 15p each, and the total receipts were £16·20. How many men and women were there at the party?*

There are two unknown quantities: the number of men and the number of women. But if the number of men is known, the number of women is 10 more.

∴ Let $x =$ the number of men.

Then $x + 10 =$ the number of women.

The facts supplied are represented as follows:

$20x =$ the money paid by the men, in pence;

$15(x + 10) =$ the money paid by the women, in pence;

$1620 =$ the total amount paid, in pence.

The equation which connects these is consequently:

$$20x + 15(x + 10) = 1620.$$

∴ $\qquad 20x + 15x + 150 = 1620$

$$35x = 1620 - 150$$

$$35x = 1470.$$

∴ $\qquad\qquad\qquad x = 42.$

∴ the number of women is $x + 10 = 52$.

∴ the solution is **42 men, 52 women**.

Example 2.　*A motorist travels from town A to town B at an average speed of 64 km/h. On his return journey his average speed is 80 km/h. He takes 9 hours for the double journey (not including stops). How far is it from A to B?*

The unknown quantity is **distance** from A to B.

Let $x =$ distance in kilometres.

Now distance = speed × time.

∴ $\qquad \dfrac{\text{distance}}{\text{speed}} = \text{time}.$

Then time for 1st journey $= \dfrac{x}{64}.$

　　,,　　　,,　　2nd　　,,　　$= \dfrac{x}{80}.$

But total time is 9 hours.

∴ the equation is

$$\frac{x}{64} + \frac{x}{80} = 9.$$

To clear of fractions multiply throughout by 320.

∴ $\qquad \left(\dfrac{x}{64} \times 320\right) + \left(\dfrac{x}{80} \times 320\right) = 9 \times 320$

$$5x + 4x = 9 \times 320$$

∴ $\qquad\qquad\qquad 9x = 9 \times 320$

and $\qquad\qquad\qquad x = 320 \text{ km}.$

This should be checked by applying the conditions in the question.

Example 3. *The weekly wages of a man and woman engaged on the same kind of work were £4·80 and £3·70 respectively. It was agreed to increase the two wages by the same amount so that the man's wage was $\frac{6}{5}$ of the woman's wage. What increase was given?*

The unknown quantity is the money to be added to the wages.

Let $x =$ the amount of increase in pence.

Then $(480 + x)$ pence $=$ the man's new wages.

$(370 + x)$,, $=$ the woman's new wages.

Then by the data

$$480 + x = \tfrac{6}{5}(370 + x).$$

Clearing fractions

$$5(480 + x) = 6(370 + x).$$
$$\therefore \quad 2400 + 5x = 2220 + 6x.$$
$$\therefore \quad 2400 - 2220 = 6x - 5x$$

and $\qquad\qquad 180 = x.$

$\therefore$ the weekly increase is 180p or £1·80.

This should be checked by adding it to each of the weekly wages and ascertaining if one is $\frac{6}{5}$ of the other.

Exercise 11.

1. From three times a certain number, n, 6 is subtracted. The result is equal to twice the number together with 6. What is the value of n?

2. There is a number such that when it is multiplied by 5 and then 14 is subtracted, the result is 348·5. Find the number.

3. From 5 times a certain number 189 is subtracted, and the remainder is one half the original number. What is that number?

4. One-third of a number added to four-fifths of itself is equal to 17. What is the number?

5. When 9 is subtracted from 6 times a certain number, the result is 45 more than twice the number. Find the number.

6. The sum of three consecutive odd numbers is 69. What are the numbers?

7. A man walks from one town to another at an average speed of 2·5 km/h. On the return he quickens his average speed to 3 km/h. The time taken for the double journey was 7 h 20 min. How far are the two towns apart?

8. The sum of a number and 4 per cent. of itself is 41·6. What is the number?

9. The perimeter of a rectangle is 44 cm. If one of the two adjacent sides is 1·8 cm longer than the other, what are the lengths of the sides?

10. Some men agree to pay equally for the use of a boat, and each pays 15p. If there had been two more men in the party, each would have paid 10p. How many men were there and how much was the hire of the boat?

11. A man distributes £2 among 20 boys, giving 5p each to some and 25p to the rest. How many boys received 25p each?

12. A man is four times as old as his son. In four years' time he will be three times as old. What are their ages now?

13. The connection between the degrees on the Celsius and Fahrenheit thermometers is that $x° C = \left(\dfrac{9x}{5} + 32 \right) °F$. What number of degrees Celsius is equivalent to 86° F?

14. A bookseller buys 120 volumes of a certain series of books. He sells some at the published price of 18p each and disposes of the remainder during a sale at 12p each. If his total receipts are £19·20, find how many volumes were sold at each price.

15. A bus is carrying 32 passengers, some with 3p tickets and the remainder with 5p tickets. If the total receipts from these passengers are £1·14, find the number of 3p fares.

CHAPTER VI

FORMULAE

48. Practical importance of formulae.

One of the most important applications of elementary Algebra is to the use of formulae. In every form of applied science and mathematics, such as mechanical engineering, electrical engineering, aeroplane construction, etc., formulae are constantly employed, and their interpretation and manipulation are essential.

49. Treatment of formulae.

In the first chapter of this book some very easy examples of formulae were introduced. With the assistance of a greater knowledge of algebraical symbols and operations and with increasing skill in their use, the student can now proceed to more difficult types.

Formulae involve three operations:

(1) Construction; (2) manipulation; (3) evaluation.

The construction of formulae cannot be indicated by any specified rules or methods. A knowledge of the principles of Algebra and skill in their application are necessary. But in general the student is concerned with formulae which have already been evolved. What he needs is skill in using them; and as his knowledge of Algebra increases, so he becomes better equipped for dealing with new examples.

The manipulation and evaluation of formulae are closely associated. A formula may need to be re-arranged or simplified before any substitution of values may be made. Experience alone will guide the student as to what manipulation is desirable in order to reach a form which is the most suitable for evaluation or some other purpose. Clear arrangement of working is always essential for accuracy.

50. Worked examples.

Example 1. *Find a formula for the total area (A) of the surface of a square pyramid as in Fig. 10 when AB = a and OQ = d.*

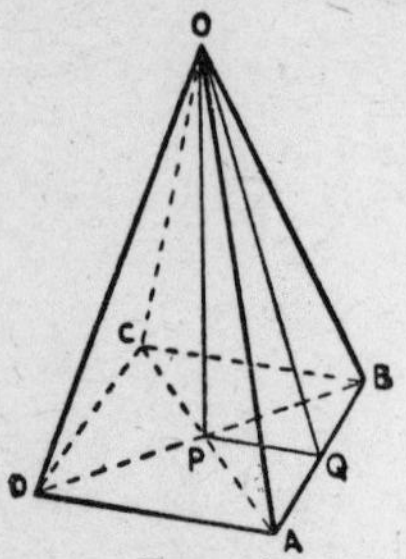

Fig. 10.

OQ is perpendicular to AB, and represents the height of the $\triangle AOB$.

The total area is made up of:

(1) Area of base.
(2) Areas of the four $\triangle$s, of which AOB is one.

Area of base $= a^2$.
Area of each $\triangle = \frac{1}{2}ad$.
$\therefore$ area of all $\triangle$s $= 4 \times \frac{1}{2}ad = 2ad$.
$\therefore$ total area of pyramid $= a^2 + 2ad$

or
$$A = a(a + 2d).$$

Example 2. *If* $L = \dfrac{W(T - t)}{w} - t$,

find L when $t = 8.5$, $w = 115$, $W = 380$, $T = 28.5$.

Substituting in

$$L = \frac{W(T - t)}{w} - t.$$

$$L = \frac{380(28.5 - 8.5)}{115} - 8.5$$

$$= \frac{380 \times 20}{115} - 8.5$$

$$= 66 - 8.5 \text{ (approx.)}$$

$$= 57.5 \text{ (approx.)}.$$

Exercise 12.

1. If $s = ut + \frac{1}{2}at^2$, find s when $u = 15$, $t = 5$, $a = 8$.

2. The volume of a cone, V, is given by the formula $V = \frac{1}{3}\pi r^2 h$, where $r =$ radius of base, $h =$ height of cone. Find V when $r = 3.5$, $h = 12$, $\pi = \frac{22}{7}$.

3. The volume of a sphere is given by the formula $V = \frac{4}{3}\pi r^3$, where $r =$ radius. Find V when $r = 3$, $\pi = \frac{22}{7}$.

4. If $E = \dfrac{Wv^2}{2g}$, find E when $W = 15 \cdot 5$, $v = 18 \cdot 8$, and $g = 32$.

5. From the formula $C = \dfrac{E + e}{R + r}$ find C when $E = 17 \cdot 6$, $e = 1 \cdot 5$, $R = 28 \cdot 4$, $r = 2 \cdot 6$.

6. In a suspension bridge the length of the cable employed is given by the formula $L = l + \dfrac{8d^2}{3l}$, where

$\quad L = $ length of the cable,
$\quad d = $ dip of the centre of the cable,
$\quad l = $ length of the span of the bridge,

all measurements being in metres. Find L when $d = 6$, $l = 56$.

7. A formula for the loading of beams is $W = \dfrac{4kbd^2}{l}$. Find W when $k = 45$, $b = 2$, $d = \frac{1}{2}$, $l = 20$.

8. If $s = \dfrac{n(n + 1)(2n + 1)}{6}$, find s when $n = 8$.

9. The following formula is used in connection with pile driving, $L = \dfrac{Wh}{d(W + P)}$. Find L when $W = 5$, $d = 1 \cdot 5$, $P = 19$, $h = 4 \cdot 5$.

10. If $R = W(x + 3t)$, find R when $W = 210$, $x = 6 \cdot 5$, $t = 0 \cdot 04$.

11. From the formula $H = \dfrac{plAN}{33\,000}$ find H when $p = 18$, $l = 2$, $A = 80$, $N = 360$.

---

51. Transformation of formulae.

In the formulae which have been examined it will be seen that **one quantity is expressed in terms of other quantities and the formula expresses the relations between them.** Thus in the formula for the volume of a cone $(V = \frac{1}{3}\pi\, r^2 h$, Exercise 12, No. 2) this volume is expressed in terms of the height and the radius of the base.

But it may be necessary to express the height of the cone

in terms of the volume and the radius of the base. In that case we would write the formula in the form:

$$h = \frac{3V}{\pi r^2},$$

that is, the formula has been transformed.

When one quantity is expressed in terms of others, as in $V = \frac{1}{3}\pi r^2 h$, the quantity thus expressed, in this case V, is sometimes called the **subject of the formula.**

When the formula was transformed into

$$h = \frac{3V}{\pi r^2},$$

the subject of the formula is now h. This process of transformation has been termed by Prof. Sir Percy Nunn " **changing the subject of the formula** ".

The transformation of formulae often requires skill and experience in algebraical manipulation; the following examples will help to illustrate the methods to be followed.

52. Worked examples.

Example 1. *From the formula* $T = \frac{\pi f d^3}{16}$ *find:*

    (1) f in terms of the other quantities.
    (2) d ,,   ,,   ,,   ,,

From $T = \frac{\pi f d^3}{16}$,

clearing the fraction

$$16T = \pi f d^3 \quad . \quad . \quad . \quad . \quad . \quad . \quad (1)$$

or $\pi f d^3 = 16T,$

dividing throughout by πd^3

$$f = \frac{16T}{\pi d^3}.$$

From (1) dividing throughout by πf

$$d^3 = \frac{16T}{\pi f}.$$

$\therefore$ $d = \sqrt[3]{\frac{16T}{\pi f}}.$

Example 2. *Transform the formula*

$$L = l + \frac{8d^2}{3l}$$

into one which expresses d in terms of the other quantities.

$$L = l + \frac{8d^2}{3l}.$$

Clearing fractions:

$$3lL = 3l^2 + 8d^2$$

or
$$8d^2 = 3lL - 3l^2.$$

$$\therefore \quad d^2 = \frac{3lL - 3l^2}{8}.$$

$$\therefore \quad d = \sqrt{\frac{3lL - 3l^2}{8}}.$$

Example 3. *The velocity, V, of water flowing through a pipe, occurs in the formula,*

$$h = 0{\cdot}03 \frac{L}{D} \times \frac{V^2}{2g}.$$

Change the subject of the formula to V.

Writing the formula as

$$0{\cdot}03 \frac{L}{D} \times \frac{V^2}{2g} = h.$$

Divide both sides by $0{\cdot}03 \frac{L}{D}$,

$$\frac{V^2}{2g} = h \div \frac{0{\cdot}03L}{D}$$

$$= h \times \frac{D}{0{\cdot}03L}$$

$$= \frac{hD}{0{\cdot}03L}.$$

$$\therefore \quad V^2 = \frac{2ghD}{0{\cdot}03L}.$$

$$\therefore \quad V = \sqrt{\frac{2ghD}{0{\cdot}03L}}.$$

Example 4. *If a — b = x(c — nd) find n in terms of the other letters.*

Attention should be fixed on the term containing n, viz. nd. This we try to isolate from the other terms.

∴ divide both sides by x.

Then
$$\frac{a - b}{x} = c - nd.$$

Transferring c

$$\frac{a - b}{x} - c = - nd.$$

∴
$$nd = c - \frac{a - b}{x}.$$

∴
$$n = \frac{1}{d}\left\{c - \frac{a - b}{x}\right\}.$$

Example 5. *The time of vibration of a simple pendulum is given by the formula*

$$t = 2\pi\sqrt{\frac{l}{g}}.$$

Find l in terms of the other quantities.

$$t = 2\pi\sqrt{\frac{l}{g}}.$$

Square both sides.

Then
$$t^2 = 4\pi^2 \times \frac{l}{g}.$$

∴
$$t^2 g = 4\pi^2 l$$

or
$$4\pi^2 l = g t^2.$$

∴
$$l = \frac{g t^2}{4\pi^2}.$$

Exercise 13.

1. The formula for the area (A) of a circle, in terms of its radius (r) is $A = \pi r^2$. Change the subject of the formula to express r in terms of the area.

2. Transform the formula for the volume of a sphere— viz. $V = \frac{4}{3}\pi r^3$ (see Exercise 12, No. 3)—into a formula in which r is expressed in terms of the volume.

3. Change the formula for the volume of a cone—viz., $V = \frac{1}{3}\pi r^2 h$—to a formula in which the subject is r.

4. The horse power of a motor is given by the formula $H = \dfrac{EC}{825}$. Express this as a formula for C.

5. The lifting force of an electro-magnet is given by the formula $F = \dfrac{B^2 A}{112 \times 10^5}$, where F is the force. Transform this into a formula of which the subject is B.

6. The amount of sag, d, in a beam under certain conditions is given by the formula $d = \dfrac{Wl^3}{48EI}$. Express this as a formula expressing l in terms of the other quantities.

7. If $v^2 = u^2 + 2as$, express s in terms of u, v, and a. Find the value of s when $u = 15$, $v = 20$, and $a = 5$.

8. There is an electrical formula $I = \dfrac{V}{R}$. Express this (1) as a formula for V and (2) as a formula for R. Find I if $V = 2$ and $R = 20$.

9. If $n^2 r + 1 = NR$, rearrange the expression so that it becomes a formula for n. Find the value of n when $N = 25$, $R = 2$, $r = 0 \cdot 81$.

10. The relation of the volume (v) of a mass of gas to the pressure (p) on it is given by the law $pv = k$.

In a certain experiment when $p = 84$, $v = 12$. Find the value of k and then express the formula giving v in terms of p and the value of k.

---

53. Literal equations.

The operations employed in changing the subject of a formula are the same in principle as those used in the solution of equations. One essential difference from the equations dealt with in Chapter V is that whereas these were concerned with obtaining numerical values when solving the equations in the formula the quantity which is the subject of the formula is expressed in terms of other quantities, and its numerical value is not determined, except when the numerical values of these quantities are known.

It is frequently necessary, however, to solve equations in which the values of the unknown quantities will be found in terms of letters which occur in the equation. Such equations are termed **literal equations**. The methods of solution are the same in principle as those employed in Chapter V. They are illustrated in the following examples.

54. Worked examples.

Example I. *Solve the equation $5x - a = 2x - b$.*

As pointed out previously (§ 46), x is understood as standing for the unknown quantity and the use of the letters a and b marks the difference between this kind of equation and those of Chapter V. The methods by which the equation is solved are the same, however, a and b being treated in the same way as ordinary numerals.

In
$$5x - a = 2x - b$$

transferring $2x$ and $-a$ respectively from one side to the other, changing the signs in so doing,

$$5x - 2x = a - b.$$

$\therefore$
$$3x = a - b$$

and
$$x = \frac{a - b}{3}.$$

Example 2. *Solve for x*
$$a(x - 2) = 5x - (a + b).$$

Removing brackets
$$ax - 2a = 5x - a - b.$$

Transferring
$$ax - 5x = 2a - a - b$$

or
$$ax - 5x = a - b.$$

This introduces a point of difference from numerical equations. With the latter we add the terms involving x by adding their coefficients. The addition in this case cannot, however, be made arithmetically. Algebraically the sum of the coefficients of x is

$$a + (-5) \text{ or } a - 5.$$

$\therefore$ we write
$$(a - 5)x = a - b.$$

Dividing both sides by the coefficient of x

$$x = \frac{a - b}{a - 5}.$$

Exercise 14.

Solve the following equations for x:

1. $5x - 4a = 0$.
2. $5x - 3a = 7a$.
3. $8x - p = 3x + 4p$.
4. $3x + 2b = 2(x + 3b)$.
5. $ax + b = 3a - b$.
6. $b(x - p) = c$.
7. $2a - b = b - bx$.
8. $3(ax - 2) + 25b = 6b$.
9. $p(x - q) = x(p - q)$.
10. $\dfrac{ax}{2} - 4 = \dfrac{b}{3} - 6$.
11. $ax - 4b = bx - b$.
12. $\dfrac{x}{a} + \dfrac{x}{3a} = 4 - x$.
13. $a(x - a) = b(x + b)$.
14. $\dfrac{x}{m} - \dfrac{1}{n} = \dfrac{3x}{mn}$.

CHAPTER VII

SIMULTANEOUS EQUATIONS

55. Simple equations with two unknown quantities.

The simple equations considered in Chapter V contained only one unknown quantity whose value it was required to determine. But many of the formulae quoted in the previous chapter contain several quantities. Cases may therefore occur in which it will be required to find the value of more than one of these. Similarly problems arise which, for their solution, involve the determination of more than one unknown.

56. Solution of simultaneous equations.

A simple problem will serve to illustrate the above statement. Suppose we are told

The sum of two numbers is 10; *what are the numbers?*

Let the two numbers be represented by x and y.

Then we know that

$$x + y = 10.$$

It is evident that there is an infinite number of solutions of this equation, such as (1, 9), (2, 8), (3·5, 6·5), etc.

If the equation be written in the form:

$$x = 10 - y.$$

This gives x in terms of y.

Whatever value be given to y in this equation, a corresponding value of x can be found, and each pair of values furnishes a solution of the equation.

If a second condition has to be fulfilled we can determine which of these pairs satisfies it.

For example, if, in addition to the statement that the sum of the numbers is 10, we are also told that one of them is four times the other; then there is only one set of the pairs of values referred to above which will satisfy both the conditions.

For if $\qquad$ $x + y = 10$ . . . . . (1)
and $\qquad$ $x = 4y$ . . . . . (2)

substituting for x in equation (1) we get:

$$4y + y = 10$$
$$5y = 10$$
$$y = 2,$$

and since $\qquad$ $x = 4y,$
then $\qquad$ $x = 8.$

∴ the solution which *satisfies both equations simultaneously is* $x = 8$, $y = 2$, and clearly there is no other solution.

For this reason such equations are called **simultaneous equations.**

It is evident that *if there are two unknown quantities whose values are required, it must be possible to form two separate equations connecting them.*

The methods employed in solving these equations are shown in the following examples.

57. Worked examples.

Example 1. *Solve the equations:*

$$2x + y = 21 \quad . \quad . \quad . \quad . \quad (1)$$
$$3x + 4y = 44 \quad . \quad . \quad . \quad . \quad (2)$$

In the method employed in this example we begin by obtaining **one letter in terms of the other.** The more convenient one is chosen, and in this case from equation (1) $2x + y = 21$, we get:

$$y = 21 - 2x \quad . \quad . \quad . \quad . \quad (3)$$

We could have found x in terms of y, but this would involve fractions and is not so convenient.

Substituting in equation (2) the value of y thus obtained from equation (1)

$$3x + 4(21 - 2x) = 44.$$

Thus we reach a simple equation with one unknown. This is solved as previously:

$$3x + 84 - 8x = 44$$
$$3x - 8x = 44 - 84$$
$$- 5x = - 40$$

or

$$5x = 40$$

and

$$x = 8.$$

Substituting this value of x in equation (3) we can find y.

Thus $y = 21 - (2 \times 8).$

∴ $y = 5.$

∴ the solution is $x = 8$, $y = 5$.

These values should be checked by substitution in *both* of the given equations.

Example 2.   In the following example a second method is shown which can frequently be employed to advantage.
Solve the equations :

$$x + y = 15 \quad . \quad . \quad . \quad . \quad (1)$$
$$3x - y = 21 \quad . \quad . \quad . \quad . \quad (2)$$

It will be seen that if the left sides of the two equations were added, y would be eliminated, since we would get $(+ y) + (- y) = 0$. Thus only x would remain.

It is clear also that the sum of the two left sides of the equations must equal the sum of the two right sides, *i.e.*,

$$(x + y) + (3x - y) = 15 + 21,$$

whence $x + 3x = 36$

$$4x = 36.$$

∴ $x = 9.$

Substituting this value of x in equation (2) or, if easier, in equation (1)

$$(3 \times 9) - y = 21.$$

∴ $27 - y = 21$

$$- y = - 6.$$

∴ $y = 6.$

∴ the solution is $x = 9$, $y = 6$.

Example 3.   In the following example both of the above methods are employed.
Solve the equations :

$$2x + 3y = 42 \quad . \quad . \quad . \quad . \quad (1)$$
$$5x - y = 20 \quad . \quad . \quad . \quad . \quad (2)$$

1st method. Substitution.

From (2) $\qquad -y = 20 - 5x.$

$\therefore \qquad\qquad\qquad y = 5x - 20 \quad . \quad . \quad . \quad . \quad$ (3)

Substituting in (1):

$$2x + 3(5x - 20) = 42$$
$$2x + 15x - 60 = 42.$$

$\therefore$

$$17x = 102$$

and $\qquad\qquad\qquad x = 6.$

Substituting for x in (3):

$$y = (5 \times 6) - 20$$

and $\qquad\qquad\qquad y = 10.$

$\therefore$ the solution is $x = 6$, $y = 10$.

2nd method. Elimination.

In this example neither letter can be eliminated by addition of the left sides of the equation, as in example 2. But by multiplying both sides of equation (2) by 3, y can be eliminated.

We proceed thus:

$$2x + 3y = 42 \quad . \quad . \quad . \quad . \quad (1)$$
$$5x - y = 20 \quad . \quad . \quad . \quad . \quad (2)$$

Multiplying by 3 throughout in (2)

$$15x - 3y = 60 \quad . \quad . \quad . \quad . \quad (3)$$

Adding (1) and (3)

$$(2x + 3y) + (15x - 3y) = 42 + 60$$

$\therefore$

$$17x = 102$$

and $\qquad\qquad\qquad x = 6.$

From this y can be found as before.

Note.—x could have been eliminated from (1) and (2) as follows:

    (*a*) Multiply throughout in (1) by 5.
    (*b*)   ,,   ,,   (2) by 2.

Then we get: $\qquad 10x + 15y = 210$
$$\qquad\qquad\qquad\quad 10x - 2y = 40.$$

Subtracting

$$(10x + 15y) - (10x - 2y) = 210 - 40.$$

$\therefore \qquad\quad 10x + 15y - 10x + 2y = 170$

whence $\qquad\qquad\qquad 17y = 170$

and $\qquad\qquad\qquad\qquad y = 10.$

Comparison of the methods. Of these two methods, that of substitution is the sounder and more general. In practice equations are seldom easily dealt with by the elimination method.

Example 4. *Find values of R_1 and R_2 which will satisfy the equations :*

$$0 \cdot 5R_1 + 1 \cdot 2R_2 = 1 \cdot 486 \quad . \quad . \quad . \quad . \quad (1)$$
$$4 \cdot 5R_1 - 2R_2 = 4 \cdot 67 \quad . \quad . \quad . \quad . \quad (2)$$

From (1)

$$0 \cdot 5R_1 = 1 \cdot 486 - 1 \cdot 2R_2.$$
$$\therefore \qquad R_1 = \frac{1 \cdot 486 - 1 \cdot 2R_2}{0 \cdot 5}$$
$$= 2 \cdot 972 - 2 \cdot 4R_2.$$

Substituting in (2):

$$4 \cdot 5(2 \cdot 972 - 2 \cdot 4R_2) - 2R_2 = 4 \cdot 67$$
$$13 \cdot 374 - 10 \cdot 8R_2 - 2R_2 = 4 \cdot 67$$
$$\therefore \qquad - 12 \cdot 8R_2 = - 8 \cdot 704$$
$$R_2 = \frac{8 \cdot 704}{12 \cdot 8}$$
$$\therefore \qquad R_2 = 0 \cdot 68.$$

Substituting in (2):

$$4 \cdot 5R_1 - (2 \times 0 \cdot 68) = 4 \cdot 67$$
$$4 \cdot 5R_1 - 1 \cdot 36 = 4 \cdot 67$$
$$4 \cdot 5R_1 = 6 \cdot 03.$$
$$\therefore \qquad R_1 = \frac{6 \cdot 03}{4 \cdot 5}$$
$$= 1 \cdot 34.$$

$\therefore$ the solution is $R_1 = 1 \cdot 34$, $R_2 = 0 \cdot 68$.

Exercise 15.

Solve the following equations:

1. $y = 2x.$
   $3x + 2y = 21.$
2. $y = 3x - 7.$
   $5x - 3y = 1.$
3. $x = 5y - 3.$
   $3x - 8y = 12.$
4. $x - y = 5.$
   $4x - y = 2x + 13.$
5. $3x - 2y = 7.$
   $x + 2y = 5.$
6. $2x - y = 3.$
   $x + 2y = 14.$

7. $2x - y = 10.$
   $3x + 2y = 29.$

8. $2(x - 4) = 3(y - 3).$
   $y - 2x = -13.$

9. $3x - 2(y + 3) = 2.$

   $2(x - 3) + 4 = 3y - 5.$

10. $\dfrac{y}{2} - x = 2.$

   $6x - \dfrac{3y}{2} = 3.$

11. $\dfrac{x}{2} - \dfrac{y}{5} = 1.$

   $y - \dfrac{x}{3} = 8.$

12. $\dfrac{a}{2} + b = 1.$

   $3a - \dfrac{b}{3} = \dfrac{31}{2}.$

13. $\dfrac{b}{3} - 2a = 5.$

   $3a + 4b = 6.$

14. $4(1 - p) = 7q + 8p.$

   $6p + q + 8 = 0.$

15. $7x + 3(y - 3) = 5(x + y).$
    $7(x - 1) - 6y = 5(x - y).$

16. $\dfrac{x}{2} - \dfrac{y}{3} = \dfrac{1}{6}.$

   $\dfrac{y}{2} - \dfrac{x}{6} = 5.$

17. $2(3a - b) = 5(a - 2).$

   $3(a + 4b) = 2(b - 3).$

18. $0.1x + 0.2y = -0.2.$
    $1.5x - 0.4y = 10.6.$

19. $1.25x - 0.75y = 1.$
    $0.25x + 1.25y = 17.$

20. $2.5x + 3.7y = 13.365.$

    $8.2x - 1.5y = 7.02.$

21. $\dfrac{x - 1}{3} + \dfrac{y + 2}{2} = 3.$

    $\dfrac{1 - x}{6} - \dfrac{y - 4}{2} = \dfrac{1}{2}.$

22. $2P - 5Q = 2.$
    $3P + 10Q = 8.6.$

23. $3(x - y) - \dfrac{1}{3}(x + y) = 30.$

    $x + y + \dfrac{5}{3}(x - y) = 22.$

24. $\dfrac{x}{3} + \dfrac{y}{2} = \dfrac{2x}{3} - \dfrac{y}{6} = 7.$

---

58. Problems leading to simultaneous equations.

Many problems require for their solution the determination of two unknown quantities. The general method of solution is similar to that employed when there is one unknown, but with the important difference that when there are two unknowns to be determined, **two equations must be formed from the data.**

The following examples illustrate the methods employed.

59. Worked examples.

Example 1. *There are two numbers such that the sum of the first and three times the second is 53, while the difference between 4 times the first and twice the second is 2. Find the numbers.*

Let x = one number.

  ,, y = the second.

Then from the first set of facts

$$x + 3y = 53 \quad . \quad . \quad . \quad . \quad (1)$$

From the second

$$4x - 2y = 2 \quad . \quad . \quad . \quad . \quad (2)$$

From (1) $x = 53 - 3y.$

Substituting in (2):

$$4(53 - 3y) - 2y = 2.$$

$\therefore$ $212 - 12y - 2y = 2.$

Collecting $-14y = -210.$

$\therefore$ $y = \dfrac{-210}{-14}$

and $y = 15.$

Substituting in $x = 53 - 3y$

                $x = 53 - (3 \times 15).$

$\therefore$ $x = 8.$

$\therefore$ **the numbers are 8 and 15.**

Example 2. *In the equation $y = mx + b$ it is known that the equation is satisfied by two pairs of values of x and y—viz. when* $x = 4, y = 6,$

and   ,, $x = 2\cdot4, y = 4\cdot5.$

What are the values of m and b?

This is an example of an important practical problem. It means that there is a law connecting x and y, the law involving m and b, which are constants. These constants must be determined before the law can be stated.

They are therefore the unknown numbers in this case. The equations connecting them are obtained by substituting the given pairs of values of x and y.

(1) When $x = 4$, $y = 6$.   ∴ on substitution

$$6 = 4m + b \quad . \quad . \quad . \quad . \quad . \quad (1)$$

(2) When $x = 2 \cdot 4$, $y = 4 \cdot 5$.

∴ $4 \cdot 5 = 2 \cdot 4m + b \quad . \quad . \quad . \quad . \quad (2)$

These are to be solved simultaneously for a and b. It is clearly a case for using the method of elimination.

Subtracting (2) from (1):

$$6 - 4 \cdot 5 = (4m + b) - (2 \cdot 4m + b)$$

or $6 - 4 \cdot 5 = 4m + b - 2 \cdot 4m - b.$

∴ $1 \cdot 5 = 1 \cdot 6m.$

∴ $m = \dfrac{1 \cdot 5}{1 \cdot 6}$ or $\dfrac{15}{16}.$

Substituting in (1)

$$\left(4 \times \frac{15}{16} \right) + b = 6.$$

∴ $b = 6 - \dfrac{15}{4}$

and $b = \dfrac{9}{4}.$

∴ the solution is $m = \dfrac{15}{16}, \ b = \dfrac{9}{4}.$

Substituting these in

$$y = mx + b$$

we get $y = \dfrac{15}{16}x + \dfrac{9}{4}$

or clearing fractions $16y = 15x + 36.$

Example 3. *A bookseller has a number of books the published price of which is 25p. After selling a certain number at this price he sells the remainder at 20p each, and his total receipts were £11. If the numbers sold at the two prices were reversed, he would have received £11·50. How many books had he in all and how many were originally sold at 25p?*

Let $x =$ the number originally sold at 25p;

 „ $y =$ „   „   „ 20p.

The amounts received for these were $25x$ pence and $20y$ pence and their total value was £11 or 1100 pence.

∴ $25x + 20y = 1100.$

When the numbers are reversed he receives $20x$ and $25y$ pence and their total value is now £11·50 or 1150 pence.

$$\therefore \qquad 20x + 25y = 1150$$

$\therefore$ the equations to be solved simultaneously are:

$$25x + 20y = 1100 \quad . \quad . \quad . \quad (1)$$
$$20x + 25y = 1150 \quad . \quad . \quad . \quad (2)$$

From (1) $\qquad\qquad 25x = 1100 - 20y$

and $\qquad\qquad\qquad x = \dfrac{1100 - 20y}{25}$

Substituting in (2):

$$\frac{20(1100 - 20y)}{25} + 25y = 1150$$

$$\therefore \qquad 4(220 - 4y) + 25y = 1150$$
and $\qquad\quad 880 - 16y + 25y = 1150$
$$\therefore \qquad\qquad\qquad\quad 9y = 270$$
$$y = 30$$

Since $\qquad x = \dfrac{1100 - 20y}{25}$

on substitution $\quad x = \dfrac{1100 - 600}{25} = \dfrac{500}{25}$

$$\therefore \qquad\qquad x = 20.$$

$\therefore$ the total number of books sold was $30 + 20 = 50$, and the number originally sold at 25p was **20**.

Exercise 16.

1. There are two numbers, x and y, such that the sum of $2x$ and y is 34, while the sum of x and $2y$ is 32. What are the numbers?

2. There are two numbers such that if to 3 times the first, twice the other is added, the sum is 72. Also if from 5 times the first number, 3 times the other is subtracted, the result is 44. What are the numbers?

3. One number is greater by 6 than twice another number, but 3 times the smaller number exceeds the greater by one. Find the numbers.

4. If from twice the greater of two numbers 17 is subtracted, the result is half the other number. If from half the greater number 1 is subtracted, the result is two thirds of the smaller number. What are the numbers?

5. In the equation $y = mx + b$, when $x = 3$, $y = 3$ and when $x = 5$, $y = 7$. Find the values of m and b and write down the equation. Then find y when $x = 6$.

6. Two quantities P and Q are connected by the formula:

$$P = \frac{m}{Q} + b.$$

When $Q = 5$, $P = 14$ and when $Q = 2$, $P = 20$. Find m and b.

7. The force (E) applied to a machine and the resistance (R) to be overcome are connected by the law

$$E = a + bR.$$

It is found that when $E = 3.5$, $R = 5$ and when $E = 5.3$, $R = 8$. Find a and b. Then find E when $R = 8$.

8. It is known that $y = ax^2 + bx^3$; when $x = 2$, $y = 5.6$, and when $x = 3$, $y = 25$. Find the values of a and b.

9. The perimeter of a rectangular lawn is 32 m. It is reduced in size so that the length is four-fifths and the breadth is three-fourths of the original dimensions. The perimeter is then 25 m. What were the original length and breadth?

10. The bill for the telephone for a quarter can be expressed in the form

$$C = a + \frac{nb}{100}$$

where C is the total cost in pounds, a is a fixed charge, n the number of calls and b the price of each call in pence. When the number of call is 104, the bill came to £5·83, and when the number was 67 the bill was £5·09. Find the fixed charge and the cost of each call.

11. The cost of 4 ties and 6 pairs of socks was £3·40, while that of 5 ties and 8 pairs of socks was £4·37. What were the prices of a tie and a pair of socks respectively?

12. The formula $S = ut + \frac{1}{2}at^2$ gives the distance (S) passed over by a moving body in t s.

In 4 s the body moves   88 m.
  ,, 6   ,,   ,,   168 m.

Find the values of u and a and then find how far the body moves in 5 s.

CHAPTER VIII

GRAPHICAL REPRESENTATION OF QUANTITIES

60. The object of graphical work.

The graphical or pictorial representation of statistics and the comparison of magnitude by means of various graphical devices are familiar features of modern life. Rows and columns of figures or groups of large or very small numbers are not always readily grasped.

Accordingly, the use of lines or columns or other figures drawn to scale which appeal to the brain through the eye, is found to be an effective method of enabling many people to realise, not only the quantities themselves but the deductions to be drawn from them.

The graphical method, developed on scientific and mathematical lines, is also largely employed in mathematics and science to illustrate certain important underlying principles. The following examples, arranged progressively, will assist the student to understand the various forms of graphical representation.

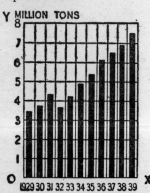

FIG. 11.

61. The column graph.

The first example, Fig. 11, is a reproduction of an actual graphical advertisement, issued by the Cement Makers' Federation to show the variations in the deliveries of cement over a period of 11 years. It is an example of what is called a column graph.

Along the straight line *OX* equal spaces are marked off which indicate, in succession, the eleven years from 1929 to 1939 inclusive.

In the spaces so formed are

constructed a series of columns of the same width and whose height, on the scale chosen, represents the number of tons of cement manufactured in each year. In order that these heights can be easily read a scale is marked along the OY, perpendicular to OX, in which each unit represents a million tons.

This assemblage of columns conveys very effectively, in a general way, the rise and fall of the production of cement during the eleven years. It also makes it possible to ascertain at a glance the approximate amount of cement produced in any one of the years.

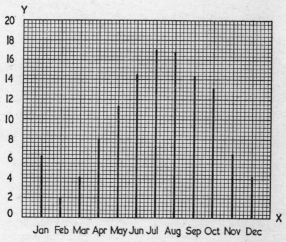

Fig. 12.

62. A straight-line graph.

The next method illustrated is more commonly used than that of the column graph. The following table shows the mean temperature for each month throughout the year.

(1969) Month	Jan.	Feb.	Mar.	Apr.	May	June	July	Aug.	Sept.	Oct.	Nov.	Dec.
Mean temp. in °C	6·2	1·7	4·1	7·9	11·6	14·3	17·0	16·6	14·6	13·3	6·2	4·2

These are exhibited graphically in Fig. 12, and the method is as follows. Draw a straight line OX, and at suitable equal intervals mark points corresponding to the twelve months.

Draw OY at right angles to OX, and with a suitable scale mark points corresponding to temperatures from 0° to 20°.

To save time and secure accuracy, specially ruled paper, called " graph paper ", is employed.

At each of the 12 points on OX draw lines perpendicular to OY to represent the corresponding temperature as indicated on the scale on OY.

These straight lines take the place of the columns in Fig. 11.

This diagram shows at a glance not only the mean temperature for each month, but gives a clear picture of the rise and fall of average temperature throughout the year, the highest and the lowest.

63. A graph.

It will be seen that when graph paper is used it is not necessary to draw the lines perpendicular to OX (as was done in Fig. 12). A point will mark the top of such a line, if it were drawn.

We now join these points by straight lines as shown in Fig. 13. The succession of lines thus formed is called a graph.

This is much more useful than the series of perpendicular lines in Fig. 12, and lends itself to important developments, as will be seen in later examples. Not only does it show more vividly the rise or fall each month, the highest and lowest temperatures, etc., but other features are apparent. For example, we get an idea of the **average rate** at which temperature is rising or falling during each month. This is indicated by the slope of the line corresponding to the month. Between November and December the slope of the line is very slight, the drop in temperature being small.

Between March and April the steep slope shows a rapid increase in temperature. The month of the sharpest rise is April, when the slope is greatest. The slope is reversed

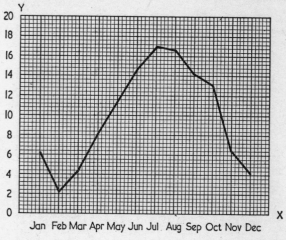

FIG. 13.

when the temperature falls, and the month when the fall in temperature is greatest is evidently January. This is rather odd and January seems to have been an exceptionally warm month in 1969; the oddness of the temperature is reflected in the oddness of the graph, which kinks at this point.

64. Examples of graphs and their uses.

Graphs are employed in almost every branch of knowledge, and a full treatment of them is impossible in a volume of this size. The following typical elementary graphs may serve to illustrate their nature and their uses. The first example is from Insurance.

Example. *The annual premiums charged by an Insurance Company for a policy of £100, at various ages, are as follows:*

Age :	25.	30.	35.	40.	45.	50.
Premium in £s. .	2·33	2·59	2·91	3·31	3·81	4·53

The amounts of the premiums have been taken to the nearest new penny. The method of "plotting" these values, as it is called, is as follows.

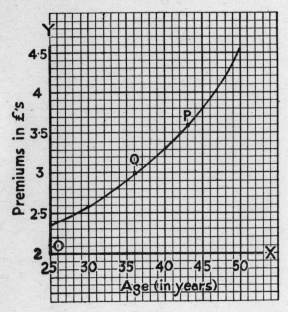

Fig. 14.

Two straight lines, *OX*, *OY*, are drawn at right angles (Fig. 14). Ages are marked off on *OX*, one small division representing a year. These are named as shown. Premiums are marked on *OY*, one small division representing 10p.

The lines OX and OY are called axes, *OX* being called the axis of *x*, and *OY* the axis of *y*, for reasons which will be apparent at later stages in the graphical work.

As there is no premium less than £2, we begin marking the premiums by placing the mark for £2 at O; similarly on the x axis we begin the ages with twenty-five years at O. By this means space on the paper is used to best advantage.

In succession we find the points which show the premium corresponding to the ages given. These are marked with dots. The student should take squared paper and get these points for himself. The process of obtaining the points is called "*plotting the points*".

When this is done it will be seen that the points *appear* to lie on a regular, smooth curve, and this has been drawn on the figure. An important point now arises.

If we are correct in assuming that points corresponding to the premiums for every fifth year lie on a regular smooth curve, then the values of these would seem to have been calculated according to a definite law or formula. If that is the case, then the premiums for the intervening years, which are not given in the table, would be calculated according to the same formula. It is therefore a reasonable deduction that the points corresponding to these, if plotted, would lie on the curve already obtained.

Accordingly, if we note the intersection of the curve with the perpendicular line corresponding to any intervening year, the position of this point with reference to the scale on the y axis indicates the amount of the premium for the year selected.

Thus, considering the point P, corresponding to the forty-third year, the premium as shown on the y axis is £3·60. Conversely, when the premium is £3 at Q, the corresponding age is thirty-six.

This process of obtaining values from the graph which lie between those plotted is called interpolation.

65. An example from Electricity.

The table below shows the resistances in ohms for given lengths of wire of the same material and cross-section.

Length in metres . .	100	120	170	220
Resistance in ohms. .	2·5	3	4·25	5·5

Draw the graph which shows the relation between resistances and length and find the resistance for a length of 200 m.

Two straight lines OX and OY are drawn as before (Fig. 15).

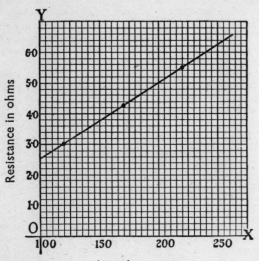

Fig. 15.

Lengths are marked along OX, beginning with 100 m at O, and each small division representing 5 m.

Resistances are marked along OY, five divisions representing 1 ohm and

∴ each division representing 0·2 ohm.

Using corresponding values from the table, points are plotted as before.

It will be seen that these points lie on a straight line.

As in the preceding example, we deduce that a definite law connects resistance and length.

Using the method of interpolation we see that *when the length is 200 m the resistance is 5 ohms.*

The graph being a straight line, it is easily produced as

in the figure. This enables us to find the resistances corresponding to lengths **beyond** those given.

For example, when the length is 250 m the resistance is a little greater than 6·2 ohms.

This is called **extrapolation**.

66. An example from Mechanics.

The following table shows the distances passed over in the times indicated by a body starting from rest. Draw a graph to show the relation between time and distance.

Time in seconds .	0	1	2	3	4	5
Distance in metres	0	2	8	18	32	50

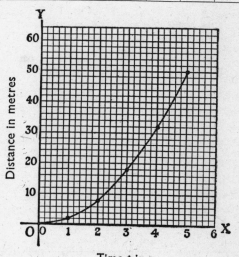

Fig. 16.

Drawing the axes *OX* and *OY*, units are selected in accordance with the range of numbers to be represented (see Fig. 16).

The points plotted are indicated by small dots. They

appear to lie on a regular curve, and this is drawn to include all the points given in the table.

As in previous examples, the regularity of the curve suggests that time and distance are connected by a definite formula. This will be familiar to those students who have studied Mechanics.

The curve is known as a **parabola**. Reasoning as in the previous example we may use the curve for interpolation. The following examples illustrate this.

Example 1. Find the distance passed over in 3·6 s.

From the curve this appears to be 26 m. The actual distance by formula is 25·92 s.

Example 2. How long will the body take to travel 42 m?

Finding the point which marks 42 m on the y axis, and then looking for the corresponding point on the curve, this is seen, from the scale on the x axis, to represent 4·6 s.

The Mechanics formula gives 4·58 s.

Exercise 17.

1. The following table gives the values of exported manufactured goods of a certain type in years as specified.

Year:	1960.	1961.	1962.	1963.	1964.	1965.
Value in million £s.	6·0	5·2	4·9	5·1	4·8	5·6

Show the variations by means of a graph.

2. The temperatures between 1000 h and 2000 h taken every two hours were as follows:

Time	1000	1200	1400	1600	1800	2000
Temp °C	8	9·1	11·9	13·7	12·1	9·0

Show these in a graph and find the probable temperature at 1300 h.

3. The expectation of life in years for males and females in this country at various ages is shown in the following table:

Age:	20.	30.	40.	50.	60.	70.
Expectation (Males) .	47·2	38·5	29·8	21·8	14·7	8·8
Expectation (Females)	50·3	41·6	32·8	24·4	16·7	10·2

From your graph estimate the expectation of life for males and females at (1) 35 years, (2) 55 years.

4. The average mass of boys of different ages is given in the following tables. Draw a graph to illustrate them.

Age (years):	11.	12.	13.	14.	15.
Mass (kg) .	40	42·5	46	50·5	57

From the graph we find:

(1) The probable average mass of a boy of 12½ years.
(2) The probable average mass at 16 years.
(3) If a boy of 13½ years has mass 46·5 kg, about how much is he below average?

5. The populations of Salford CB and Poole MB at certain times is shown in the table below. Draw, using the same axes and scales, graphs showing the change in the two populations. (1) From your graphs, when are the populations likely to be equal? (2) What will this equal population be?

Draw the graph showing the difference in the two populations. (3) When does this show the difference as being zero? [N.B. In order to use the same axes and scales as before, draw the graph of " difference + 100 000 ": the " zero-line " will then be 100 000.]

Population in thousands

Year:	1951	1961	1966	1968	1969
Salford . .	177	154	144	140	138
Poole . .	83	92	97	99	102

6. A train starts from rest, and its speed at intervals during the first minute is given in the following table:

Time (s) .	0	5	10	20	30	40	50	60
Speed (m/s) .	0	8·5	14·6	23	29·2	33·6	37	39

Draw a graph showing the relation between time and speed. Does it appear that a definite law connects these? What is your estimate of the speed after 23 s?

7. The following table gives the annual premium £P payable during life by a man aged A years next birthday for a whole life assurance of £1000. Draw the graph.

A .	20	25	30	35	40	45	50	55
P .	11·60	13·10	13·30	17·50	21	25·50	31·50	40

A mistake was made in working out these figures: find from your graph which entry is wrong and estimate the correct entry. Estimate the annual premium for men aged 24, 42, 53 next birthday.

8. The following table shows the sales of gas for public lighting in the North West in the years 1963/4–1968/9. What is the likely sale in 1969/70?

Gas sales in kilotherms

Year	63/4	64/5	65/6	66/7	67/8	68/9
Gas .	1933	1466	1017	723	552	434

9. If $y = \frac{1}{2}(x - 3)(x + 1)$, work out values of y for successive whole number values of x from $x = -3$ to $x = 5$. Draw a graph of $y = \frac{1}{2}(x - 3)(x + 1)$. From your graph, read off the two values of x that make y equal to 1.

CHAPTER IX

THE LAW OF A STRAIGHT LINE; CO-ORDINATES

67. The straight-line graph.

Among the graphs considered in the previous chapter were some which were straight lines. This regular arrangement of the plotted points suggested that a definite law governed the relations between the quantities, corresponding values of which were the basis of the plotting. This is an important principle which calls for further investigation.

The following example will serve as a starting-point.
The proprietor of a restaurant calculated the following figures showing the connection between his net profits and the number of customers. Exhibit the connection by a graph.

Number of Customers .	240	270	300	350	380
Net profit in £s. . .	0·50	2·00	3·50	6·00	7·50

Note.—" Net profit " means total receipts from customers, less expenses.

Let x = the number of customers.

,, y = the net profit.

Choosing two axes as usual, values of x will be plotted on the x axis, values of y on the y axis.

When plotted as in Fig. 17, the points are found to lie on a straight line.

Interpolation and extrapolation can be used as previously, but two important questions arise:

(1) *What is the number of customers when there is no profit?*

Zero on the profit (y) scale is shown where the graph cuts the x axis. Producing the graph it cuts at the point (A), where $x = 230$, *i.e.*,

when $x = 230, y = 0.$

(2) *What happens when the number of customers is less than* 230?

If there is no profit when the number of customers is 230, there will be a loss when the number is below 230.

Let the straight line be produced below the *x* axis, as in Fig. 17. The amount of the loss, as with the profit, will be shown on the *y* axis, which must be produced.

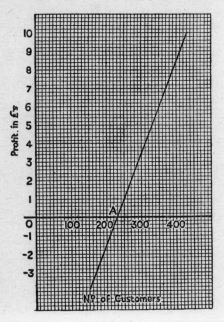

FIG. 17.

The question of indicating loss as contrasted with profit now arises as it did in Chapter IV. We must proceed in the same way. The loss will be marked with negative numbers on the scale, as shown. It will then be seen that on the *y* axis there is thus constructed part of a complete number scale (see § 36).

From this scale, it appears that when x (the number of customers) is 180, y (the loss) is £2·50.

Problems will also arise when it will be necessary to use negative values of x, on the x axis produced. Consequently for a complete graph we shall require

> complete number scales on both axes, with zero common to both.

In the particular example above, a negative number of customers would have no intelligible meaning.

68. The law represented by a straight-line graph.

Our next step is to discover the nature of the law which a straight-line graph represents and how it can be formulated. We will use the above problem as our example. In that problem it was stated that the net profit was equal to the total receipts less the expenses. But the total receipts equal (the number of customers) × (the average amount paid by each).

Let £a = the average amount thus paid.

Then since

x = the number of customers,

ax = the total amount paid in pounds.

Let £b = the expenses.

Then net profits are $(ax - b)$ pounds, *i.e.*,

$$y = ax - b.$$

This gives the value of y in terms of a, x, and b, and is the form of the law connecting them.

Of the four letters, a and b remain unchanged, while the number of customers (x) and consequently the profit (y) vary.

∴ the law is not completely stated until we know the values of a and b.

The method of doing this is suggested by example 2, § 59.

Two pairs of corresponding values of x and y can be obtained from the table of values or from the graph.

For example, when

$$x = 320,\ y = 4·50$$
when $\qquad x = 250,\ y = 1·00.$

Substituting these in

$$y = ax - b$$
$$4{\cdot}5 = 320a - b \quad . \quad . \quad . \quad . \quad (1)$$
$$1 = 250 - b \quad . \quad . \quad . \quad . \quad (2)$$

Subtracting (2) from (1)

$$3{\cdot}5 = 70a$$

and $\qquad\qquad a = 0{\cdot}05.$

Substituting in (1) we get

$$b = 11{\cdot}5.$$

∴ the equation is

$$y = 0{\cdot}05x - 11{\cdot}5.$$

This is the law of the straight line.

If it is correct it must be satisfied by any corresponding pair of values of x and y, and this should be tested by the student.

In particular if $y = 0$, *i.e.* there is no profit

$$0{\cdot}05x - 11{\cdot}5 = 0$$

whence $\qquad\qquad 0{\cdot}05x = 11{\cdot}5$

and $\qquad\qquad\qquad x = 230.$

This agrees with the result found above.

Thus it can be demonstrated that the equation

$$y = 0{\cdot}05x - 11{\cdot}5$$

is satisfied by the co-ordinates of any point on the line and so is called the equation to the straight line.

69. Graph of an equation of the first degree.

The equation which is represented by the straight line in the above problem (viz., $y = 0{\cdot}05x - 11{\cdot}5$) is of the first degree, that is, it contains no higher powers than the first of x and y.

Two questions now suggest themselves:

(1) Can every equation of the first degree in x and y be represented graphically by a straight line?

(2) Conversely, can every straight line be represented algebraically by an equation of the first degree?

The answers to these questions will be apparent later, but for the present we will confine ourselves to drawing the graphs of some typical equations of the first degree. From these the answer to the first question may be deduced.

Examples are given to illustrate the methods to be employed. In all of these, since both positive and negative values of x and y will be involved, complete number scales will be used on both axes.

Before calculating corresponding values of x and y, the student who is not familiar with the work should revise § 56.

70. Worked examples.

Example I. *Draw the graph of the equation:*

$$2y - 4x = 3.$$

This is not in the form used above, but it can readily be transformed thus:

$$2y - 4x = 3$$
$$2y = 4x + 3.$$
$$\therefore \qquad y = 2x + 1\cdot5.$$

Giving suitable values to x, the corresponding values of y can be calculated, and so we get the following table:

x .	−2	−1	0	1	1·8	2
y .	−2·5	−0·5	1·5	3·5	5·1	5·5

A straight line is fixed by two points, but in drawing it from its equation three points should always be taken to check accuracy. In this example, as we are verifying the truth of a proposition, a number of points are taken, so as to make it clear that all such points lie on the straight line. The graph appears as shown in Fig. 18.

This could be checked by finding for any point on the line the corresponding values of x and y. These values should satisfy the equation.

Intercepts.

It should he noted that when

(1) $x = 0, y = 1\cdot5,$

when (2) $y = 0, x = -0\cdot75.$

(1) 1·5 Is called the intercept on the y axis ($x = 0$).
(2) —0·75   ,,   ,,   ,,   x   ,,   ($y = 0$).

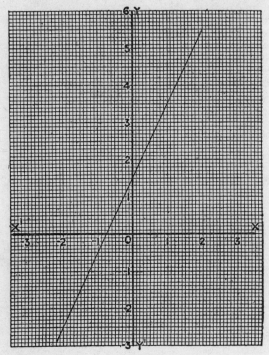

FIG. 18.

Example 2.   *Draw the graph of the equation:*

$$2x + y = 1.$$

Transforming the equation we get:

$$y = -2x + 1 \quad . \quad . \quad . \quad . \quad (A)$$

A table of corresponding values can be constructed as follows:

x .	-2	-1	0	1	2	3
y .	5	3	1	-1	-3	-5

With these values the straight line shown in Fig. 19 is drawn.

For **intercepts on the axes**

when $x = 0$, $y = 1$ (intercept on y axis)

when $y = 0$, $x = 0 \cdot 5$ (intercept on x axis).

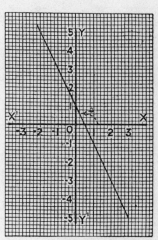

FIG. 19.

It should be noted from this example that when the coefficient of x in the equation arranged as in (A) is negative, the angle which the straight line makes with the x axis in an anti-clockwise direction is greater than a right angle. When the coefficient is positive, as in Fig. 18, this angle is less than a right angle.

Exercise 18.

Draw the graphs which are represented by the following equations. In each case find the intercepts on (1) the y axis, (2) the x axis.

1. $y = x + 2$.
2. $y = 1 \cdot 5x - 1$.
3. $4y = 6x - 5$.
4. $3x + 2y = 6$.
5. $\dfrac{x}{3} + \dfrac{y}{2} = 2$.
6. $2(x - 3) = 4(y - 1)$.

7. The equation $y = ax + b$ is satisfied by the following pairs of values of x and y:

(1) $x = 1$, $y = 5$.
(2) $x = 2$, $y = 7$.

Find the values of a and b, and substitute in the equation. Draw the graph of this equation and find its intercept on the y axis.

8. A straight line makes an intercept of 2 on the y axis and of 4 on the x axis. Find its equation.

---

71. Position in a plane ; co-ordinates.

When a point is " *plotted* ", as in graphs previously considered, its position on the graph was fixed by the corresponding values of x and y which were given in the table. For example, let P be a point such that in the table

Fig. 20.

$x = 2$, $y = 3$. When plotted, the point appears as in Fig. 20, where:

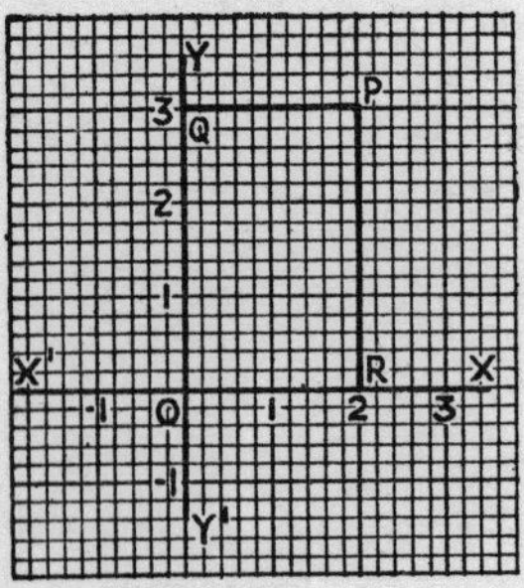

PQ is 2 units in length and parallel to OX,

and

PR is 3 units in length and parallel to OY.

The intersection of these two straight lines fixes the position not only in the graph, but also relative to the axes OX and OY. The position of any other point can be similarly determined when its distances from the two axes are known.

By such means might a man discover where, in a field, he had hidden a treasure—an act all too necessary in parts of Europe in these times. By remembering the distances OR and OQ along two boundaries of the field, he can " plot the point P " in the field. A little reflection will convince the student that the boundaries need not be at right angles to one another.

The distances PQ and PR (Fig. 20) which thus fix the position of a point are called the **co-ordinates** of P with respect to the two axes.

PQ, parallel to the x axis, is called the x co-ordinate (or the abscissa).

PR, parallel to the y axis, is called the y co-ordinate (or the ordinate).

The notation employed to denote co-ordinates is (2, 3) or in general (x, y). The x co-ordinate is always placed first inside the bracket.

Thus (5, 2) would represent the point in which $x = 5$, $y = 2$. In this way the positions of points relative to two axes can be described concisely. If, however, all points are to be included, complete number scales must be used on both axes, as shown in Fig. 21. In this way four divi-

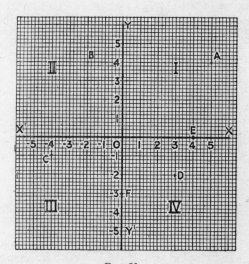

FIG. 21.

sions, called quadrants, are found in the plane; numbered I, II, III, IV as shown.

The signs of the co-ordinates are regulated by the positive and negative parts of the number scales.

The rule of signs is as follows:

If (x, y) be the co-ordinates of a point:

> x is +ve when measured to the right of 0.
> x is —ve   ,,   ,,   ,,   left of 0.
> y is +ve   ,,   ,,   up from 0.
> y is —ve   ,,   ,,   down from 0.

As examples the co-ordinates of A, B, C, D in the figure are:

A is $(5, 4)$, B is $(-2, 4)$, C is $(-4, -1)$, D is $(3, -2)$.

Points on the axes.

If a point is on the x axis the y co-ordinate is zero. Thus E is $(4, 0)$.

If a point is on the y axis the x co-ordinate is zero. Thus F is $(0, -3)$.

The intersection of the axes, O, is called the origin. Since it lies on both axes, its co-ordinates are $(0, 0)$.

Thus the **position of every point on a plane relative to two axes can be determined by co-ordinates.**

Latitude and longitude are a practical example of use of co-ordinates. They describe the position of a place with reference to the equator and the meridian through Greenwich as axes.

The introduction of co-ordinates was due to Descartes, who published his book on Analytical Geometry in 1638.

Exercise 19.

1. Write down the co-ordinates of the points in Fig. 22 marked $A, B, C \ldots G$.

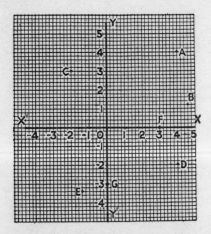

Fig. 22.

2. Join OA and FC in Fig. 22, and find the co-ordinates of their point of intersection.

3. Plot the points $(3, 1)$, $(1, 3)$, $(0, 3)$, $(0, -3)$, $(4, 2\cdot6)$, $(-2, 1)$, $(-4, -2)$, $(3, 0)$, $(-2, 0)$.

4. Draw the straight lines joining the points $(3, 1)$ and $(1, 3)$ and also $(-2, 1)$ and $(4, 2\cdot6)$ as plotted in the previous question. What are the co-ordinates of the point of intersection of the two lines?

5. Plot the points $(-3, 2)$, $(0, 2)$, $(2, 2)$, $(4, 2)$. What do you notice about these points?

6. Plot the points $(3, 3)$, $(1, 1)$, $(-1, -1)$, $(-2, -2)$. What do you notice about these points?

7. Draw a straight line through $(3, 0)$ parallel to YOY'. What do you notice about the co-ordinates of points on this line?

---

72. A straight line as a locus.

Let A (Fig. 23) be a point such that its co-ordinates are equal.

Let $(x_1 y_1)$ be its co-ordinates.

Then $$y_1 = x_1.$$

Join OA and draw AP perpendicular to OX.

Then $$AP = OP.$$

∴ by Geometry $\angle OAP = \angle AOP$

since $\llcorner APO$ is a right angle.

∴ each of the angles OAP, AOP is $45°$.

∴ A lies on a straight line passing through the origin and making $45°$ with the axis OX.

Let B be any other point with equal co-ordinates (x_2, y_2) so that $BQ = OQ$, i.e., $y_2 = x_2$.

Then, for the same reasons as above, B also lies on a straight line passing through the origin and making $45°$ with the x axis.

This must be the straight line OA, since only one straight line can pass through the origin and make an angle of $45°$ with the x axis.

Similarly all other points with equal co-ordinates lie on the same straight line—i.e., **the straight line OA, produced, is the locus of all points with equal co-ordinates.**

These co-ordinates all satisfy the equation $y = x$, which is therefore the equation of the line.

This equation can be written in the form

$$\frac{y}{x} = 1.$$

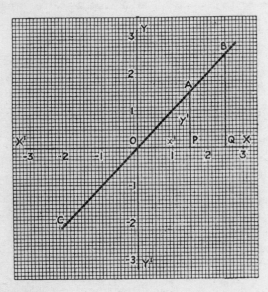

Fig. 23.

$\therefore$ for any point such as A,

$$\frac{AP}{OP} = 1.$$

This ratio is constant for all points on the straight line, and is called the **gradient** of the line.

A similar result holds for every straight line, consequently **a straight line is a graph which has a constant gradient**.

To include all points with equal co-ordinates the straight line must be produced into the opposite (3rd) quadrant.

Then for any point on this line the co-ordinates are equal,

but they are both negative. Thus for the point C (Fig. 23) these co-ordinates are $(-2, -2)$ and the gradient is $\dfrac{-2}{-2}$, *i.e.*, unity.

The student who has learnt sufficient Geometry, or Trigonometry, will know, without the above demonstration, that the ratio of $\dfrac{AP}{OP}$ is constant for any point A. See also § 159.

73. Equation of any straight line passing through the origin.

The conclusions reached above apply equally to all straight lines through the origin. The lines differ only in the gradient—*i.e.*, in the value of $\dfrac{y}{x}$.

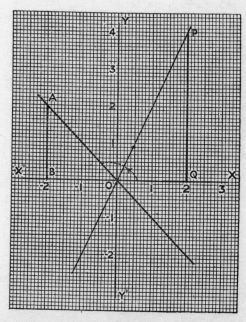

Fig. 24.

For example, if the gradient is 2, then $\dfrac{y}{x} = 2$ and $y = 2x$.

This is shown in Fig. 24. For any point P, on the line OP, is

the ratio $$\frac{PQ}{OQ} = \frac{4}{2} = 2.$$

Generally, if the gradient be denoted by m we have

$$\frac{y}{x} = m.$$
$$\therefore \qquad y = mx.$$

This is the general form of the equation of any straight line through the origin, where m *denotes the gradient.*

Negative gradient. If m is negative, the line will pass through the 2nd and 4th quadrants.

Thus if $m = -1$ the equation is

$$y = -x.$$

The straight line is shown in Fig. 24.

Considering any point A, the gradient is

$$\frac{AB}{OB} = \frac{+2}{-2} = -1.$$

It should be noted that the angle made with the axis of x is 135°, angles being always measured in an anti-clockwise direction.

74. Graphs of straight lines not passing through the origin.

In Fig. 25 the straight line AOB is the graph of $y = x$.

If we now plot the graph of $y = x + 2$, it is evident that for any value of x the value of y in $y = x + 2$ is greater by 2 than the corresponding value of y in $y = x$.

$\therefore$ the line for $y = x + 2$ must be parallel to $y = x$, but each point is two units higher in the y scale. Thus in Fig. 25 the point A is raised to A', the origin to D, B to B', etc.

$A'DB'$ therefore represents the graph of $y = x + 2$.

The straight line $y = x + 2$ is the locus of all points whose co-ordinates are such that the y co-ordinate = the x co-ordinate $+ 2$. It has the same gradient as $y = x$, but its intercept on the axis of y is $+ 2$.

Similarly the line $y = x - 3$ is parallel to the line $y = x$, with each point on it lowered by 3 units in the y scale.

Evidently then we can generalise and state that $y = x + b$ will always represent a straight line parallel to $y = x$ and with an intercept of b units on the y axis for any value of b.

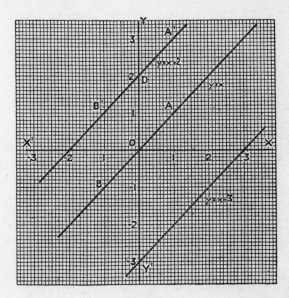

FIG. 25.

The same conclusions hold for lines with different gradients. For example, the equation $y = 2x + b$ will always represent a straight line parallel to $y = 2x$, *i.e.*, having the same gradient, and with an intercept of b units on the y axis.

Generalising, let $m =$ the gradient of a straight line.

Then $y = mx + b$ always represents a straight line parallel to $y = mx$, *i.e.*, with the same gradient, and with an intercept of b units on the y axis.

Examples.

(1) $y = 4x - 7$ is a straight line whose gradient is 4, and whose intercept on the y axis is $- 7$.

(2) $y = 0·05x - 11·5$ (see § 68) represents a straight line of gradient 0·05 and intercept on the y axis equal to $- 11·5$.

As shown in § 70, every equation of the first degree in two unknowns can be reduced to the form $y = mx + b$.

We therefore conclude that the graph of every equation of the first degree in two unknowns is a straight line.

Further, it is evident that **the equation is satisfied by the co-ordinates of any point on the straight line.**

75. Graphical solution of simultaneous equations.

The conclusions reached in the previous paragraph can be used, as shown in the following example, to solve simultaneous equations of the first degree.

Example. *Solve the equations:*

$$x + 2y = 5 \quad . \quad . \quad . \quad . \quad (1)$$
$$3x - 2y = 7 \quad . \quad . \quad . \quad . \quad (2)$$

(1) *Draw the graph of* $x + 2y = 5$. The table of co-ordinates is as follows:

x .	0	-1	2	5
y .	2·5	3	1·5	0

Note.—When $x = 0$, y is the **intercept on the x axis;** when $y = 0$, x is the **intercept on the x axis.** It is useful to obtain these two points.

The graph is the straight line marked A in Fig. 26.

(2) *Draw the graph of* $3x - 2y = 7$. The table of values is:

x .	0	1	$\frac{7}{3}$	4
y .	$-3·5$	-2	0	2·5

The graph is the **straight line marked B.**

Applying the conclusions, reached in § 74.

Line *A* contains all those points whose co-ordinates satisfy the equation $x + 2y = 5$.

Line *B* contains all those points whose co-ordinates satisfy the equation $3x - 2y = 7$.

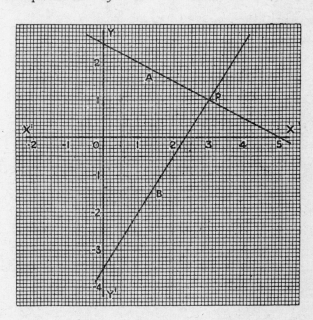

FIG. 26.

There is one point, and one point only, whose co-ordinates satisfy both equations.

That point is *P*, the intersection of the two graphs.

The co-ordinates of *P*, by inspection, are (3, 1).

∴ the solution of the equation is

$$x = 3, y = 1.$$

The student should compare the above conclusions with the algebraical treatment in § 55.

Exercise 20.

1. With the same axes draw the graphs of:

    (1) $y = x.$ (2) $y = 2x.$ (3) $y = \frac{1}{2}x.$
    (4) $y = -x.$ (5) $y = -2x.$

2. With the same axes draw the graphs of:

    (1) $y = x.$ (2) $y = x + 1.$ (3) $y = x + 3.$
    (4) $y = x - 1.$

3. With the same axes draw the graphs of:

    (1) $y = \frac{1}{2}x.$ (2) $y = \frac{1}{2}x + 2.$ (3) $y = \frac{1}{2}x - 1.$

4. With the same axes draw the graphs of:

    (1) $y = x + 2.$ (2) $y = 2x + 2.$
    (3) $y = -x + 2.$ (4) $y = \frac{1}{2}x + 2.$

5. Draw the graphs of:

    (1) $2x + y = 3.$ (2) $x - 2y = 4.$
    (3) $5x + 2y = 10.$ (4) $4x - 5y = 10.$

In each case find the intercepts on the x axis and on the y axis.

6. Solve graphically the following pairs of equations and check by algebraical solutions:

    (1) $x + y = 7.$ (2) $2x + y = 7.$
        $x - y = 1.$ $3x - 5y = 4.$
    (3) $2x + 3y = 2.$
        $4x + y = -6.$

7. The straight line whose equation is $y = ax - 1$ passes through the point $(2, 5)$. What is the value of a? What is the intercept on the y axis?

8. The straight line whose equation is $y = 2x + b$ passes through the point $(1, 3)$. What is the value of b? Draw the straight line. What is its intercept on the y axis?

9. The points $(1, 1)$ and $(2, 4)$ lie on the straight line whose equation is $y = ax + b$. Find a and b and write down the equation. What is the intercept of the straight line on the y axis?

(*Hint.*—See example 2 of § 59.)

10. Draw the straight line $y = 2x + 1$.

Draw the straight line parallel to it and passing through the point $(0, 3)$. What is the equation of this line?

11. Find the equation of the straight line passing through the points $(1, 3)$ and $\left(2, \dfrac{11}{3}\right)$. What is the gradient of the line?

12. Find the equation of the straight line passing through the points $(2, 4)$ and $(3, 5)$. Find its gradient.

CHAPTER X

MULTIPLICATION OF ALGEBRAICAL EXPRESSIONS

76. (1) **When one factor consists of one term.**

THIS has been considered in § 25, when it was shown that

$$x(a + b) = xa + xb.$$

77. (2) **Product of binomial expressions**—*i.e.*, with two terms.

A typical example is $(x + a)(y + b)$.

As in § 25, a geometrical illustration will help to make clear what is the product and how it is obtained.

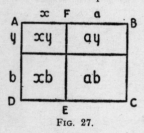

FIG. 27.

In Fig. 27 *ABCD* is a rectangle with the sides $(x + a)$ and $(y + b)$ units of length, and divided to represent x, a, y, and b units of length. Lines are drawn parallel to the sides dividing the whole rectangle into smaller ones whose areas represent the products xy, xb, ay and ab by means of their areas.

The area of the whole rectangle

$$= (x + a)(y + b) \text{ sq. units.}$$

Also the area of the whole rectangle

$$= \text{area of } ADEF + \text{area of } EFBC$$
$$= x(y + b) + a(y + b).$$

$$\therefore \quad (x + a)(y + b) = x(y + b) + a(y + b)$$
$$= xy + bx + ay + ab.$$

This grouping suggests the method of multiplying $(x + a)(y + b)$ algebraically.

The second binomial $(y + b)$ is multiplied in turn by each

term of the first factor. The sum of these is the final product.

As the order in multiplication is immaterial so far as the final product is concerned, this could also have been obtained as follows. Writing the factors in the reverse order:

$$(y + b)(x + a) = y(x + a) + b(x + a)$$
$$= xy + ay + bx + ab.$$

This is illustrated in Fig. 27.

Worked examples.

$$(a + b)(c + d) = a(c + d) + b(c + d)$$
$$= ac + ad + bc + bd.$$
$$(x + 5)(y + 3) = x(y + 3) + 5(y + 3)$$
$$= xy + 3x + 5y + 15.$$
$$(a + b)(c - d) = a(c - d) + b(c - d)$$
$$= ac - ad + bc - bd.$$
$$(a - x)(b - y) = a(b - y) - x(b - y)$$
$$= ab - ay - bx + xy.$$
$$(x - 4)(y + 3) = x(y + 3) - 4(y + 3)$$
$$= xy + 3x - 4y - 12.$$

If the first terms in each factor are alike, the same method is followed, the product being simplified afterwards if necessary. Thus:

$$(x + a)(x + b) = x(x + b) + a(x + b)$$
$$= x^2 + bx + ax + ab.$$

This could be expressed as

$$x^2 + (b + a)x + ab.$$

This result suggests a quick way of obtaining the product mentally. The coefficient of x in the answer is the **sum** of a and b. The last term is their **product**.

When a and b are numbers the sum and product will be evaluated, and the expression simplified.

Examples.

$$(x + 6)(x + 5) = x(x + 5) + 6(x + 5)$$
$$= x^2 + 5x + 6x + 30$$
$$= x^2 + 11x + 30.$$

$$(a + 9)(a + 4) = a^2 + (9 + 4)a + 9 \times 4$$
$$= a^2 + 13a + 36.$$
$$(x + 2)(x - 7) = x(x - 7) + 2(x - 7)$$
$$= x^2 - 7x + 2x - 14$$
$$= x^2 + x(-7 + 2) - 14$$
$$= x^2 - 5x - 14.$$
$$(a - 8)(a - 3) = a(a - 3) - 8(a - 3)$$
$$= a^2 - 3a - 8a + 24$$
$$= a^2 - 11a + 24.$$
$$(x - 8)(x - 2) = x^2 + (-8 - 2)x + (-8) \times (-2)$$
$$= x^2 - 10x + 16.$$
$$(x - 8)(x + 2) = x^2 + (-8 + 2)x + (-8 \times 2)$$
$$= x^2 - 6x - 16.$$
$$(x + 8)(x - 2) = x^2 + (+8 - 2)x + \{+8 \times (-2)\}$$
$$= x^2 + 6x - 16.$$

78. When the coefficients of the first terms are not unity the rule still holds. Thus:

$$(px + a)(qx + b) = px(qx + b) + a(qx + b)$$
$$= pqx^2 + pbx + aqx + ab,$$

which may be written as

$$pqx^2 + (pb + aq)x + ab.$$

The form of the coefficients of x in the last line should be noted. It will be used later.

Numerical examples of this form are common. The following illustrations show how to apply the rule quickly.

Worked examples.

Example I. $(2x + 5)(3x + 4)$

$$= 2x(3x + 4) + 5(3x + 4)$$
$$= 6x^2 + x(2 \times 4 + 5 \times 3) + 20$$
$$= 6x^2 + 23x + 20.$$

The coefficient of the middle term can be obtained by multiplying as shown by the arrows below; then add the results:

$$(2x + 5)(3x + 4).$$

Example 2. $(3x + 7)(2x + 1)$.

Product $= 6x^2 + x\{(3 \times 1) + (7 \times 2)\} + 7$
$= 6x^2 + 17x + 7$.

Example 3. $(7x - 5)(2x + 3)$.

Product $= 14x^2 + x\{(7 \times 3) - (5 \times 2)\} - 15$
$= 14x^2 + 11x - 15$.

Example 4. $(3x - 2)(4x - 7)$.

Product $= 12x^2 + x\{(3 \times -7) + (-2 \times 4)\} + 14$
$= 12x^2 - 29x + 14$.

If any difficulty is experienced the working should be set as shown in the second line of Example 1.

79. Multiplication of a trinomial.

The method shown in § 77 may be usefully adapted to certain cases in which one of the factors is a trinomial.

Example 1. $(x + 2)(x^2 - x + 1)$
$= x(x^2 - x + 1) + 2(x^2 - x + 1)$
$= x^3 - x^2 + x + 2x^2 - 2x + 2$
$= x^3 + x^2 - x + 2$ (on collecting like terms).

Example 2. $(a + b)(a^2 - ab + b^2)$
$= a(a^2 - ab + b^2) + b(a^2 - ab + b^2)$
$= a^3 - a^2b + ab^2 + a^2b - ab^2 + b^3$
$= a^3 + b^3$ (on collecting like terms).

Exercise 21.

Write down the following products:

1. $(a + x)(b + y)$.   2. $(c + d)(e + f)$.
3. $(ax + b)(cy + d)$.   4. $(a - x)(b - y)$.
5. $(x - y)(a - b)$.   6. $(a - x)(b + y)$.
7. $(a + x)(b - y)$.   8. $(a + 2)(b + 3)$.
9. $(a - 2)(b - 3)$.   10. $(a - 2)(b + 3)$.

11. $(a + 2)(b - 3)$. 12. $(x + 7)(x + 5)$.
13. $(ab + 6)(ab + 3)$. 14. $(x + 10)(x + 3)$.
15. $(x - 10)(x - 3)$. 16. $(x + 10)(x - 3)$.
17. $(x - 10)(x + 3)$. 18. $(p + 8)(p - 12)$.
19. $(x - 4y)(x - 8y)$. 20. $(x - 4y)(x + 8y)$.
21. $(x + 4y)(x - 8y)$. 22. $(a + 2b)(2a + 5b)$.
23. $(3x - 4y)(3x - 5y)$. 24. $(4x + 1)(7x + 2)$.
25. $(2x - 3)(3x - 1)$. 26. $(3x + 1)(3x - 4)$.
27. $(1 + 3y)(1 - 4y)$. 28. $(6x + 1)(3x - 5)$.
29. $(7x - 3y)(2x + 5y)$. 30. $(3a - 7b)(6a - 5b)$.

31. Simplify $\{(x + y)(a + b) - (ay + yb)\}$ and divide the result by x.

32. Simplify $\{(a + b)(a - c) + bc\} \div a$.

33. Find the value of $(2x - y)(x + y) - (2x + y)(x - y)$ when $x = 3$, $y = 2$.

34. Simplify $(1 - 2x)(1 + 3y) - (1 - 2y)(1 + 3x)$ and find its value when $x = 0 \cdot 1$, $y = 0 \cdot 2$.

35. Find the following products:

  (a) $(x - y)(x^2 + xy + y^2)$.
  (b) $(a + 2)(a^2 - 2a + 4)$.
  (c) $(1 + x)(1 - x + x^2)$.
  (d) $(x + a)(x^2 + 2ax + a^2)$.

---

80. Square of a binomial expression.

In § 77 it was shown that

$$(x + a)(x + b) = x^2 + x(a + b) + ab.$$

Since this is true whatever the values of the letters, let $b = a$.

Then $(x + a)(x + a) = x^2 + x(a + a) + a \times a$
$= x^2 + 2xa + a^2$.

∴ $$(x + a)^2 = x^2 + 2ax + a^2.$$

If a be replaced by $- a$ throughout, then

$$(x - a)^2 = x^2 - 2ax + a^2.$$

Geometrical illustration.

Fig. 28 shows an illustration of the above result by means of a square whose side is $x + a$.

It is a modification of Fig. 27 and will be readily understood without further explanation.

The student should draw a similar figure to illustrate

$$(x - a)^2 = x^2 - 2ax + a^2.$$

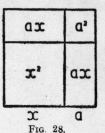

Fig. 28.

Examples.

$(x + 1)^2 = x^2 + 2x + 1.$

$(x - 1)^2 = x^2 - 2x + 1.$

$(a + 9)^2 = a^2 + 18a + 81.$

$(1 - 5xy)^2 = 1 - 10xy + 25x^2y^2.$

$(2x + 7y)^2 = 4x^2 + 28xy + 49y^2.$

$(3a - 10b)^2 = 9a^2 - 60ab + 100b^2.$

81. Square of a trinomial.

In the product

$$(x + c)^2 = x^2 + 2cx + c^2.$$

Since x may have any value, replace it by $a + b$. Then

$$(a + b + c)^2 = (a + b)^2 + 2c(a + b) + c^2$$
$$= a^2 + 2ab + b^2 + 2ac + 2bc + c^2,$$

or, re-arranging,

$$(a + b + c)^2 = a^2 + b^2 + c^2 + 2ab + 2bc + 2ac.$$

This may be stated in words as follows:

The square of a trinomial is equal to
(sum of squares of each term) + (twice the product of the three terms taken two at a time in every possible way).

Examples.

$(a + b + 1)^2 = a^2 + b^2 + 1 + 2ab + 2a + 2b.$

$(x - y + z)^2$
$\quad = x^2 + (-y)^2 + z^2 + 2(x \times -y) + 2(xz) + 2(-y \times z)$
$\quad = x^2 + y^2 + z^2 - 2xy + 2xz - 2yz.$

$(a - b - c)^2 = a^2 + b^2 + c^2 - 2ab - 2ac + 2bc.$

$(x + y - 5)^2 = x^2 + y^2 + 25 - 10x - 10y + 2xy.$

Care should be taken in applying the laws of signs.

82. Cube of a binomial.

$(a + b)^3$ may be written as $(a + b)(a + b)(a + b)$

$$= (a + b)(a^2 + 2ab + b^2).$$

Multiplying these as shown in § 79

$$(a + b)^3 = a(a^2 + 2ab + b^2) + b(a^2 + 2ab + b^2)$$
$$= a^3 + 2a^2b + ab^2 + a^2b + 2ab^2 + b^3.$$
$$\therefore (a + b)^3 = a^3 + 3a^2b + 3ab^2 + b^3 \text{ (on collecting like terms).}$$

Similarly

$$(a - b)^3 = a^3 - 3a^2b + 3ab^2 - b^3.$$

Examples.

$(x + 1)^3 = x^3 + 3x^2 + 3x + 1.$
$(x - 1)^3 = x^3 - 3x^2 + 3x - 1.$
$(1 + a)^3 = 1 + 3a + 3a^2 + a^3.$
$(1 - a)^3 = 1 - 3a + 3a^2 - a^3.$
$(2x + 3y)^3 = (2x)^3 + \{3(2x)^2 \times 3y\} + \{3(2x) \times (3y)^2\} + (3y)^3$
$\qquad = 8x^3 + 36x^2y + 54xy^2 + 27y^3.$
$(x - 3y)^3 = x^3 - 9x^2y + 27xy^2 - 27y^3.$

Note.—In the cube of $(x - y)$ the signs are alternately positive and negative.

Exercise 22.

Write down the following squares in full:

1. $(x + 2)^2$.
2. $(x - 2)^2$.
3. $(a + 3b)^2$.
4. $(a - 3b)^2$.
5. $(2x + y)^2$.
6. $(x - 2y)^2$.
7. $(ab + 10)^2$.
8. $(xy - 3)^2$.
9. $(4x + 5y)^2$.
10. $(4x - 5y)^2$.
11. $(5xy + 6)^2$.
12. $(1 - 10x^2)^2$.
13. $(5x^2 + 3y^2)^2$.
14. $(3xy - 2y^2)^2$.
15. $\left(x + \dfrac{1}{y}\right)^2$.
16. $\left(\dfrac{1}{x} - \dfrac{1}{y}\right)^2$.
17. $\left(a + \dfrac{1}{3}\right)^2$.
18. $\left(\dfrac{1}{2} - \dfrac{3y}{4}\right)^2$.
19. $\{(x + y) + 1\}^2$.
20. $\{1 - (x - 2y)\}^2$.
21. $(a + b - c)^2$.
22. $(x - y + z)^2$.
23. $(2x + 3y - 5z)^2$.
24. $(4a - 2b - 1)^2$.

Write down the following cubes in full:

25. $(x + y)^3$.

26. $(x - y)^3$.

27. $(a + 2)^3$.

28. $(a - 2)^3$.

29. $(p + q)^3$.

30. $(p - q)^3$.

31. $(2x + y)^3$.

32. $(x - 2y)^3$.

33. $(3a - 1)^3$.

34. $(1 - 3b)^3$.

35. Simplify the following: $(x + y)^2 - (x^2 + y^2)$.

36. Simplify: (1) $(a + b)^2 - (a - b)^2$.

(2) $(a - b)^2 - (a + b)^2$.

37. Simplify $(x + 10)^2 - (x - 10)^2$.

38. If $x = 3y + 1$, express $x^2 + 4x + 4$ in terms of y and find its value when $y = 1$.

39. A square lawn the side of which is x m is surrounded by a path (Fig. 29) which is a m wide. Find an expression for the area of the path in terms of x and a. What is the area of the path when $x = 30$ m and $a = 2$ m?

40. If the lawn of the previous question were a rectangle, x m by y m and the path were a m wide, find an expression for its area.

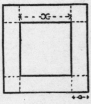

Fig. 29.

---

83. Product of sum and difference.

If a and b be any two numbers, then the product of their sum and difference is expressed by $(a + b)(a - b)$.

Using the method of § 77 to find the value of this

$$(a + b)(a - b) = a(a - b) + b(a - b)$$
$$= a^2 - ab + ab - b^2.$$

$$\therefore \ (a + b)(a - b) = a^2 - b^2$$

This important result can be expressed in words as follows:

The product of the sum and difference of two numbers is equal to the difference of their squares.

Examples.

$(x + 9)(x - 9) = x^2 - 81$.

$(ab + 10)(ab - 10) = a^2b^2 - 100$.

$(4x + 5)(4x - 5) = 16x^2 - 25.$

$(1 + x)(1 - x) = 1 - x^2.$

$(a - \frac{1}{2})(a + \frac{1}{2}) = a^2 - \frac{1}{4}.$

$\left(\dfrac{a}{3} + \dfrac{b}{4}\right)\left(\dfrac{a}{3} - \dfrac{b}{4}\right) = \dfrac{a^2}{9} - \dfrac{b^2}{16}.$

$\{(a + b) + c\}\{(a + b) - c\} = (a + b)^2 - c^2.$

$\{a + (b + c)\}\{a - (b + c)\} = a^2 - (b + c)^2.$

Exercise 23.

Write down the following products:

1. $(a + x)(a - x).$
2. $(p + q)(p - q).$
3. $(a + 2b)(a - 2b).$
4. $(4x + 3)(4x - 3).$
5. $(2x + 1)(2x - 1).$
6. $(1 + 6x)(1 - 6x).$
7. $(1 + a^2)(1 - a^2).$
8. $(2x^2 + 1)(2x^2 - 1).$
9. $(x^2 + y^2)(x^2 - y^2).$
10. $(3xy + 2)(3xy - 2).$
11. $(12xy + 1)(12xy - 1).$
12. $(\frac{1}{4}x - 7)(\frac{1}{4}x + 7).$
13. $\{(x + y) + z\}\{(x + y) - z\}.$
14. $(a + x + y)(a + x - y).$
15. $(2a + 3b + 1)(2a + 3b - 1).$
16. $(x - 2y + 6)(x - 2y - 6).$
17. $\{a + 2(b + c)\}\{a - 2(b + c)\}.$
18. $\{2x + 3(y + z)\}\{2x - 3(y + z)\}.$
19. $(x + \frac{2}{3})(x - \frac{2}{3}).$
20. $\left(\dfrac{x}{2} + \dfrac{y}{3}\right)\left(\dfrac{x}{2} - \dfrac{y}{3}\right).$
21. $(x + a + \frac{1}{2})(x + a - \frac{1}{2}).$

CHAPTER XI

FACTORS

84. THE work of this chapter will be the converse of that in the previous one. That was concerned with methods of obtaining the products of certain algebraical expressions. In this chapter we shall seek to find the factors of expressions of different types.

A converse operation, in general, is more difficult than the direct one, and so it is in this case. By rules, on the whole simple and easily applied, the products of various kinds of factors are found. But in seeking to find the factors whose product produces a given expression, rules, even when they are formulated, are often long and tedious. In the main we rely on trial methods, which are, however, not haphazard but based on those rules by which the product was obtained.

It is always possible to obtain a product when the factors are given, but we cannot always find factors for expressions. Most expressions have no factors, and so we can deal only with special types, such as were obtained as products in the work of the preceding chapter.

85. Monomial factors—*i.e.*, factors consisting of one term only.

This is the converse of the theorem stated in § 25. There we saw that

$$x(a + b) = xa + xb.$$

If we start with $xa + xb$ and wish to factorise it, we see by inspection that x **is a factor of each term.** It is therefore a factor of the whole expression. To find the other factor we divide each term by x and add the quotients. As a result we get $a + b$.

$$\therefore \qquad xa + xb = x(a + b).$$

In finding the factors by inspection we are guided by a knowledge of the process by which when $(a + b)$ is multiplied by x the result is $xa + xb$.

86. Worked examples.

Example 1. *Find the factors of* $6a^2 + 3ac$.

In this case we see that there is more than one factor common to each term.

    (1) **3** is the highest common factor of the numerical coefficients.

    (2) a is a factor of the other parts of each term.

$\therefore$ $3a$ is a factor of **each term**, and is therefore a factor of the whole expression. Dividing each term by it,

$$6a^2 + 3ac = 3a(2a + c).$$

Example 2. *Factorise* $5x^2y^2 - 10x^2y + 20y^2$.

The highest numerical factor common to each term is 5, and the only other factor is y.

$\therefore$ $5x^2y^2 - 10x^2y + 20y^2 = 5y(x^2y - 2x^2 + 4y)$.

Exercise 24.

Express the following as factors:

1. $6x + 12$.
2. $3ab + 2a$.
3. $4xy + 2y^2$.
4. $6a^2 - 4ab$.
5. $14x^2y^2 - 7xy$.
6. $16 - 32a^2$.
7. $a^2 - ab + ac$.
8. $x^3 + 3x^2 - x$.
9. $15a^3 - 5a^2b + 3a^2b^2$.
10. $6a^2c - 15ac^2$.
11. $a^2b + ab^2 - abc$.
12. $\dfrac{bc^2}{6} - \dfrac{abc}{9}$.

13. Fill in the blank in the following:

$$(7 \cdot 4 \times 13^2) + (7 \cdot 4 \times a^2) = 7 \cdot 4(\qquad).$$

14. Calculate the following as easily as you can:

$$18 \cdot 6^2 + 18 \cdot 6 \times 1 \cdot 4.$$

87. Binomial factors.

In § 77 we saw that

$$(x + a)(y + b) = x(y + b) + a(y + b)$$
$$= xy + bx + ay + ab.$$

If we require the factors of $xy + bx + ay + ab$ and similar expressions, we must work backwards through the steps shown above.

The first step is to reach the stage $x(y + b) + a(y + b)$.

To obtain this the four terms of the expression must be suitably arranged in two pairs such that:

(1) The terms in each pair have a common factor.
(2) When this common factor is taken out, the same expression must be left in each pair.

$\therefore$ we group $xy + bx + ay + ab$ in the form

$$(xy + bx) + (ay + ab).$$

Then taking out the common factor in each pair we get

$$x(y + b) + a(y + b).$$

Then $y + b$ is a factor of both parts of the expression and must therefore be a factor of the whole.

Thus we get, on taking it out,

$$(y + b)(x + a).$$

We will now apply this method to expressions of which the factors are not previously known, as they were in the example.

88. Worked examples.

Example I. *Find the factors of* $a^2 + cd + ad + ac$.

The first two terms have no common factor. Consequently the order of the terms could be changed in order to get two pairs, each with a common factor.

$\therefore$ we write $\quad a^2 + cd + ad + ac$
$$= (a^2 + ad) + (cd + ac).$$

In this arrangement
(1) a is common to the terms of the first group and c to those of the second,

(2) the same expression $a + d$ will be the other factor of each group.

$\therefore$ we get $a(a + d) + c(a + d)$.

$(a + d)$ being a factor of both parts is a factor of the whole.

$\therefore$ expression $= (a + d)(a + c)$.

Note.—Another possible arrangement was:

$$(a^2 + ac) + (ad + cd)$$
$$= a(a + c) + d(a + c)$$
$$= (a + c)(a + d).$$

There are always two possible ways of grouping.

Example 2. *Factorise, if possible,* $ab + ac + bc + bd$.

This example is given to show the beginner where he might go astray.

The expression can be grouped as $a(b + c) + b(c + d)$. But the expressions in the brackets in the two parts are **different**. It is a bad mistake, which is sometimes made, to write down the factors as $(b + c)(c + d)(a + b)$. Not one of these is a factor of the expression. Moreover their product would be an expression of the third degree. The given expression is of the second degree.

On trying different groupings, it will be found that it is not possible to arrange in two groups, having the same factor in each group.

There are no factors of this expression.

Example 3. *Find the factors of* $ab - 5a - 3b + 15$.

By arrangement into suitable pairs:

$$ab - 5a - 3b + 15$$
$$= (ab - 5a) - (3b - 15)$$
$$= a(b - 5) - 3(b - 5)$$
$$= (b - 5)(a - 3).$$

It should be carefully noted that when an expression is placed in brackets with minus sign in front, as was done with $- 3b + 15$ above, **the signs within the brackets must be changed.** This is in agreement with the reverse rule given in § 26, cases (3) and (4).

Exercise 25.

Find the factors of:

1. $ax + ay + bx + by$.
2. $pc + qc + pd + qd$.
3. $ab - bd + ae - de$.
4. $ax - cx - ay + cy$.
5. $x^2 + px + qx + pq$.
6. $x^2 - gx - hx + gh$.
7. $ab + 5b + 6a + 30$.
8. $ab - 5b - 6a + 30$.
9. $ab - 5b + 6a - 30$.
10. $2ab - 10a + 3b - 15$.
11. $ax^2 + a^2x - ab - bx$.
12. $x^2 + ax - bx - ab$.

---

89. The form $x^2 + ax + b$.

It was seen in § 77 that the product of two factors such as $(x + 6)(x + 5)$ was $x^2 + 11x + 30$.

Reversing the process, we must now consider how to find the factors of $x^2 + 11x + 30$.

In the general case it was shown that

$$(x + a)(x + b) = x^2 + x(a + b) + ab.$$

In this product

    (1) the coefficient of x is the sum of the numbers a and b,

    (2) the term independent of x, i.e., ab, is the product of these numbers.

Consequently in finding the factors of an expression such as $x^2 + 11x + 30$, we must find by trial two factors of 30 whose sum is $+ 11$. They are $+ 5$ and $+ 6$.

∴ the factors are $(x + 5)(x + 6)$.

90. Worked examples.

Example 1. *Factorise $x^2 + 13x + 36$.*

There are several pairs of factors of 36—viz. (1×36), (2×18), (3×12), (4×9) and (6×6).

We look for the pair whose sum is $+ 13$; this pair is $(9, 4)$.

∴ $x^2 + 13x + 36 = (x + 4)(x + 9)$.

Example 2. *Factorise $x^2 - 13x + 36$.*

In this case the sign of the middle term is $-$ and of the last term is $+$.

∴ we must look for two **negative** factors of 36 whose sum is $- 13$

These are $- 9$ and $- 4$.

$$\therefore \qquad x^2 - 13x + 36 = (x - 9)(x - 4).$$

Example 3. *Factorise* $y^2 - 13y + 30$.

Proceeding as in the last example, we get:

$$y^2 - 13y + 30 = (y - 10)(y - 3).$$

Example 4. *Factorise* $x^2 - 5x - 36$.

When the last term is negative **the factors of 36 must have opposite signs.**

∴ the coefficient of x is the sum of a **positive and negative** number, and the negative number must be the greater **numerically**, or we might say that 5 is the difference of the two factors numerically.

The factors are clearly $- 9$ and $+ 4$.

$$\therefore \qquad x^2 - 5x - 36 = (x - 9)(x + 4).$$

Note.—The larger of the two numbers has the same sign as the middle term.

Example 5. *Factorise* $x^2 + 12x - 28$.

The two factors of 28 which differ by 12 are 14 and 2. They are of opposite signs, and their sum is $+ 12$.

∴ the factors required are $+ 14$ and $- 2$.

$$\therefore \qquad x^2 + 12x - 28 = (x + 14)(x - 2).$$

Example 6. *Factorise* $a^2 - 8ab - 48b^2$.

The introduction of the second letter makes no difference to the method followed, but b will appear in the second term of each factor.

Thus $\quad a^2 - 8ab - 48b^2 = (a - 12b)(a + 4b).$

Exercise 26.

Find the factors of:

1. $x^2 + 3x + 2$.
2. $x^2 - 3x + 2$.
3. $x^2 + 5x + 6$.
4. $x^2 - 5x + 6$.
5. $x^2 + 7x + 6$.
6. $x^2 + 9x + 20$.
7. $x^2 - 12xy + 20y^2$.
8. $a^2 - 15ab + 36b^2$.

9. $x^2y^2 + 15xy + 54$.　　　　10. $a^2b^2 - 19ab + 48$.

11. $y^2 - 21y + 108$.　　　　　12. $x^2 - 12xy + 35y^2$.

13. $x^2 - x - 2$.　　　　　　　14. $x^2 + x - 2$.

15. $x^2 + xy - 6y^2$.　　　　　16. $x^2 - xy - 6y^2$.

17. $b^2 - 2b - 3$.　　　　　　18. $b^2 + 2b - 3$.

19. $x^2 + 13x - 48$.　　　　　20. $x^2 - 13x - 48$.

21. $x^2 - xy - 110y^2$.　　　　22. $a^2 - 11a - 12$.

23. $a^2 - a - 12$.　　　　　　24. $p^2 + p - 72$.

25. $p^2 - 34p - 72$.　　　　　26. $1 - 9x + 20x^2$.

27. $1 - 8x - 20x^2$.　　　　　28. $x^2y^2 - 3xy - 88$.

29. $p^2 + 4p - 45$.　　　　　30. $p^2 + pq - 56q^2$.

---

91. The form $ax^2 + bx + c$.

The factors of such an expression as this, where a, b, c are numbers, will be of the form which occurred in the converse operation in § 78. These factors are best obtained by trial, as indicated in the following examples.

Example 1. *Find the factors of $2x^2 + 7x + 3$.*

Write down possible pairs of factors systematically and find the middle term as shown in § 78 until the correct coefficient of x is found. In this example the possibilities are :

(1) $(2x + 3)(x + 1)$.

(2) $(2x + 1)(x + 3)$.

In (1) the coefficient of x in the product is 5.

　,, (2)　　　,,　　　　　,,　　,,　　　,, 7.

The second pair is therefore the correct one—*i.e.*,

$$2x^2 + 7x + 3 = (2x + 1)(x + 3).$$

Example 2. *Find the factors of $6x^2 + 17x - 3$.*

The minus sign of the last term indicates that the factors will have opposite signs in the second terms. This increases the number of possible pairs of factors. Among the possible pairs are:

(1) $(2x - 1)(3x + 3)$ or $(2x + 1)(3x - 3)$.

(2) $(6x - 1)(x + 3)$ or $(6x + 1)(x - 3)$.

Proceeding as shown in § 78, we find that the first pair of (2) is the one required, since the coefficient of x is

$$(6 \times 3) + (1 \times - 1) = 18 - 1 = + 17.$$
$$\therefore \qquad 6x^2 + 17x - 3 = (6x - 1)(x + 3).$$

Example 3.   *Find the factors of* $4x^2 - 17x - 15$.

The sign of the last term is again negative; therefore the factors will have opposite signs.   Among the possibilities are:

(1) $(2x + 5)(x - 3)$ or $(2x - 5)(2x + 3)$.
(2) $(4x + 5)(x - 3)$ or $(4x - 5)(x + 3)$.
(3) $(4x + 3)(x - 5)$ or $(4x - 3)(x + 5)$.

The first pair of (3) gives for the coefficient of x

$$(4 \times - 5) + (3 \times 1) = - 20 + 3 = - 17.$$
$$\therefore \qquad 4x^2 - 17x - 15 = (4x + 3)(x - 5).$$

Exercise 27.

Complete the following factors:

1. $3x^2 + 10x + 8 = (3x \qquad)(x \qquad)$.
2. $12x^2 - 17x + 6 = (4x \qquad)(3x \qquad)$.
3. $12x^2 - 28x - 5 = (6x \qquad)(2x \qquad)$.
4. $9x^2 + 43x - 10 = (9x \qquad)(x \qquad)$.

Find the factors of the following:

5. $2x^2 + 3x + 1$.   6. $3x^2 - 4x + 1$.
7. $2x^2 + 5x + 2$.   8. $6x^2 + 5x + 1$.
9. $4x^2 - 8x + 3$.   10. $5x^2 - 6x + 1$.
11. $6x^2 - 11x + 3$.   12. $12x^2 + 11x + 2$.
13. $2a^2 + a - 1$.   14. $2a^2 - a - 1$.
15. $2a^2 - a - 6$.   16. $10b^2 - b - 2$.
17. $10b^2 + b - 2$.   18. $8y^2 - 14y - 15$.
19. $12x^2 + 5x - 2$.   20. $14c^2 - 17c - 6$.

92. Squares of binomial expressions.

A trinomial expression which is the square of a binomial can be recognised by applying the rule given in § 80.

The standard forms are:

$$(a + b)^2 = a^2 + 2ab + b^2$$
and
$$(a - b)^2 = a^2 - 2ab + b^2.$$

If the trinomial is a square, the following conditions must be satisfied:

(1) The first and last terms, when the trinomial has been suitably arranged, are **exact squares** and **positive.**

(2) The middle term must be $\pm$ (twice the product of the square roots of the first and third terms).

Example 1. *Write $x^2 + 6x + 9$ as a square of a binomial expression.*

(1) The first and third terms are the squares of x and 3.

(2) The middle term is $+ (2 \times \sqrt{x^2} \times \sqrt{9}) = 6x$.

$\therefore \qquad x^2 + 6x + 9 = (x + 3)^2.$

Similarly $\qquad x^2 - 6x + 9 = (x - 3)^2.$

Example 2. *Is $4x^2 + 6x + 9$ a complete square?*

(1) The first and third terms are the squares of $2x$ and 3.

(2) For a complete square the middle term should be $+ (2 \times \sqrt{4x^2} \times \sqrt{9}) = + 12x.$

But the middle term is $+ 6x.$

$\therefore$ the expression is not a complete square.

Example 3. *Is $4x^2 - 20x + 25$ a complete square?*

The middle term is

$$- 2(\sqrt{4x^2}) \times \sqrt{25} = 2 \times 2x \times 5 = 20x.$$

$\therefore \qquad 4x^2 - 20x + 25 = (2x - 5)^2.$

93. Difference of two squares.

In § 83 it was shown that

$$(x + a)(x - a) = x^2 - a^2.$$

Conversely $\quad (x^2 - a^2) = (x + a)(x - a).$

$\therefore$ an expression which is the difference of two squares has for its factors the **sum** and the **difference** of the numbers which are squared.

Note.—No real factors can be found for the sum of two squares—*i.e.*, $x^2 + a^2.$

94. Worked examples.

Example 1. *Factorise $100x^2 - 1$.*

The numbers squared are $10x$ and 1.

∴ the factors are the sum and difference of these—*i.e.*,

$$100x^2 - 1 = (10x + 1)(10x - 1).$$

Example 2. *Factorise $36a^2b^2 - 25$.*

The numbers squared are $6ab$ and 5.

∴ $$36a^2b^2 - 25 = (6ab + 5)(6ab - 5).$$

Example 3. *Factorise $(a + b)^2 - c^2$.*

Although one of the terms squared is replaced by a binomial expression, the rule above still applies.

∴ $$\begin{aligned}
(a + b)^2 - c^2 &= \{(a + b) + c\}\{(a + b) - c\} \\
&= (a + b + c)(a + b - c).
\end{aligned}$$

Example 4. *Factorise $(a + b)^2 - (c - a)^2$.*

$$\begin{aligned}
(a + b)^2 &- (c - a)^2 \\
&= \{(a + b) + (c - a)\}\{(a + b) - (c - a)\} \\
&= (a + b + c - a)(a + b - c + a) \\
&= (b + c)(2a + b - c).
\end{aligned}$$

95. Evaluation of formulae.

The above rule is frequently employed in transforming formulae and in arithmetical calculations.

Example 1. *Find the value of $47 \cdot 5^2 - 22 \cdot 5^2$.*

$$\begin{aligned}
47 \cdot 5^2 - 22 \cdot 5^2 &= (47 \cdot 5 + 22 \cdot 5)(47 \cdot 5 - 22 \cdot 5) \\
&= 70 \times 25 \\
&= 1750.
\end{aligned}$$

Example 2. *Find the area of a ring between two concentric circles of radii 97 mm and 83 mm, respectively. The area of a circle is πr^2 where the radius is r.*

The area of the ring is the difference between the areas of the two circles.

Difference in area

$$\begin{aligned}
&= \pi \times 97^2 - \pi \times 83^2 \\
&= \pi(97^2 - 83^2) \quad \text{taking out common factor} \\
&= \pi\{(97 + 83) \times (97 - 83)\} \\
&= \pi \times 180 \times 14 \\
&= 2520\pi \text{ mm}^2.
\end{aligned}$$

Exercise 28.

Write each of the following as the square of a binomial expression:

1. $p^2 + 2pq + q^2$.
2. $x^2 - 4xy + 4y^2$.
3. $9x^2 + 6x + 1$.
4. $16x^2 - 40xy + 25y^2$.
5. $x^2 + x + \frac{1}{4}$.
6. $\frac{a^2}{9} + \frac{ab}{3} + \frac{b^2}{4}$.

Express as complete squares:

7. $(a + b)^2 + 4(a + b) + 4$.
8. $(x - y)^2 - 10(x - y) + 25$.

Find the factors of the following:

9. $x^2 - 100$.
10. $a^2b^2 - 25$.
11. $4x^2 - 9y^2$.
12. $25a^2 - 16b^2$.
13. $121x^2 - 36y^2$.
14. $144p^2 - 169q^2$.
15. $25 - 16a^2$.
16. $1 - 225x^2$.
17. $8a^2 - 50b^2$.
18. $3x^2 - 75$.
19. $5x^2 - 45y^2$.
20. $(a + b)^2 - c^2$.
21. $(x + 2y)^2 - 16z^2$.
22. $1 - \frac{81}{16}y^2$.
23. $x^2 - (y + z)^2$.
24. $a^2 - (x - 2y)^2$.
25. $(x - 8)^2 - 49$.
26. $(a + b)^2 - (a - b)^2$.

Find the numerical values of the following:

27. $65^2 - 35^2$.
28. $82^2 - 68^2$.
29. $49^2 - 39^2$.
30. $24^2 - 18^2$.
31. $4\cdot25^2 - 1\cdot75^2$.
32. $17\cdot5^2 - 12\cdot5^2$.

33. Find the value of $\pi(r_1^2 - r_2^2)$ when $\pi = \frac{22}{7}$, $r_1 = 12\cdot5$, $r_2 = 8\cdot5$.

34. If $S = \frac{v^2 - u^2}{2f}$ find S when $v = 14\cdot5$, $u = 2\cdot5$ and $f = 1\cdot5$.

35. In the formula $w = k(D^2 - d^2)$ find w, when $k = 2\cdot4$, $D = 8\cdot5$, $d = 7\cdot5$.

36. If $y = \frac{m(v^2 - u^2)}{2p}$ find y when $m = 24$, $v = 44$, $u = 16$, $p = 32$.

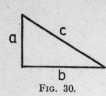

FIG. 30.

37. If $s = \frac{1}{2}gt^2$ find the difference between the values of s when $t = 8$ and $t = 6$. Take $g = 10$.

38. In the right-angled triangle (Fig. 30) a and b represent the lengths of the sides containing the right angle and c the length of the hypotenuse. We know from Geometry that

$$c^2 = a^2 + b^2.$$

(1) Find a when $c = 52$, $b = 48$.
(2) Find b when $c = 65$, $a = 25$.

---

96. Sum and differences of two cubes.

In § 79, Example 2, it was shown that

$$(a + b)(a^2 - ab + b^2) = a^3 + b^3.$$

By using the same method we get:

$$(a - b)(a^2 + ab + b^2) = a^3 - b^3.$$

These results enable us to obtain factors of the expressions:

$$a^3 + b^3 \text{ and } a^3 - b^3.$$

Re-arranging them we have:

$$a^3 + b^3 = (a + b)(a^2 - ab + b^2) \quad . \quad . \quad \text{(A)}$$
$$a^3 - b^3 = (a - b)(a^2 + ab + b^2) \quad . \quad . \quad \text{(B)}$$

It may aid the memory to note that

(1) In (A), the *sum of two cubes*, the first factor is the *sum of the numbers*.
(2) In (B), the *difference of two cubes*, the first factor is the *difference of the numbers*.

The other factors differ only in the sign of *ab*.

(1) When the first factor is the *sum* the sign is *minus*.
(2) When the first factor is the *difference* the sign is *plus*.

97. Worked examples.

$$x^3 + 1 = (x + 1)(x^2 - x + 1).$$
$$x^3 - 1 = (x - 1)(x^2 + x + 1).$$
$$1 - x^3 = (1 - x)(1 + x + x^2).$$
$$a^3 - 27 = (a - 3)(a^2 + 3a + 9).$$
$$8x^3 + 125 = (2x + 5)(4x^2 - 10x + 25).$$

Exercise 29.

Find the factors of:

1. $x^3 + c^3$.

2. $y^3 - a^3$.

3. $1 + 8a^3$.

4. $x^3 - 64$.

5. $8 + 27c^3$.

6. $R^3 - 1$.

7. $m^3 - 125n^3$.

8. $x^3y^3 + \frac{1}{8}$.

9. $\frac{1}{x^3} + \frac{1}{y^3}$.

10. $\frac{1}{x^3} - \frac{1}{y^3}$.

CHAPTER XII

FRACTIONS

98. Algebraic fractions.

IN Algebra fractions have the same fundamental meaning as in Arithmetic, and are subject to the same rules in operations with them. They differ only in the use of letters in the algebraic forms.

An **algebraic fraction** is one in which the denominator is an algebraic expression. The numerator may or may not be algebraic in form.

Thus $\dfrac{2}{a}$ is an algebraic fraction, but $\dfrac{a}{2}$ is not. The latter means $\frac{1}{2}a$, and thus $\frac{1}{2}$ is merely a fractional coefficient.

Similarly $\dfrac{a}{a+b}$ is an algebraic fraction but $\dfrac{a+b}{6}$ is not.

99. Laws of fractions.

It was stated above that algebraic fractions are subject to the same laws as arithmetical fractions. It is unnecessary to state these laws here, but a few illustrations may serve to remind the student how they operate. It was pointed out in § 24 that

$$\frac{a}{b} = \frac{a \times m}{b \times m} \text{ and } \frac{a}{b} = \frac{a \div m}{b \div m},$$

where m is any number.

By means of this rule algebraic fractions can be simplified or reduced to lowest terms by dividing both numerator and denominator by the same number. This can be expressed by the phrase " common factors are cancelled." This was assumed in § 21.

100. Reduction of fractions.

Examples were given in § 24, but further instances are now given depending on the operations of the previous chapter.

Example 1. *Simplify* $\dfrac{a + b}{a^2 - b^2}$.

Using the rule of § 93,

$$\frac{a + b}{a^2 - b^2} = \frac{a + b}{(a + b)(a - b)}$$
$$= \frac{1}{a - b} \quad \text{(on cancelling } (a + b)\text{)}.$$

Example 2. *Simplify* $\dfrac{x^2 + 4x - 12}{x^2 + x - 6}$.

Factorising numerator and denominator,

$$\frac{x^2 + 4x - 12}{x^2 + x - 6} = \frac{(x - 2)(x + 6)}{(x - 2)(x + 3)}$$
$$= \frac{x + 6}{x + 3}.$$

The mistake is sometimes made of cancelling **terms** instead of **factors**. Thus, there can be no cancelling with the 3 and 6 in the above answer. Only **factors of the whole expressions** in numerator and denominator can be cancelled.

Example 3. *Simplify* $\dfrac{3a(a^2 - 4ab + 4b^2)}{6a^2(a^2 + 3ab - 10b^2)}$.

Factorising, the fraction becomes

$$\frac{3a(a - 2b)^2}{6a^2(a + 5b)(a - 2b)}.$$

The factors $3a(a - 2b)$ are common to numerator and denominator. Cancelling them, the fraction becomes,

$$\frac{a - 2b}{2a(a + 5b)}.$$

Exercise 30.

Simplify the following fractions:

1. $\dfrac{2xyz}{6x^2z^2}$.

2. $\dfrac{15abd^3}{20a^3bd}$.

3. $\dfrac{2x^2 + 2xy}{4xy - 4y^2}.$

4. $\dfrac{3a^3 + 9ab}{6a^3b - 6a^2b^2}.$

5. $\dfrac{6x^2y + 6xy^2}{4x^3y^2 + 4x^2y^3}.$

6. $\dfrac{a^2 - b^2}{a^2 - ab}.$

7. $\dfrac{2x - 6}{x^2 - 5x + 6}.$

8. $\dfrac{x^2 - 9x + 20}{x^2 + x - 20}.$

9. $\dfrac{a^3 + 6a^2b + 9ab^2}{a^2b + 5ab^2 + 6b^3}.$

10. $\dfrac{x^3 - xy^2}{x^4 - xy^3}.$

101. Multiplication and division.

If necessary the student should revise § 24 and then proceed to the harder examples now given.

Example 1. *Simplify* $\dfrac{x}{x + 1} \div \dfrac{x^2}{x^2 - 1}.$

Factorising and inverting the second fraction:

$$\frac{x}{x + 1} \div \frac{x^2}{x^2 - 1} = \frac{x}{x + 1} \times \frac{(x + 1)(x - 1)}{x^2}.$$

Cancelling common factors we get:

$$\frac{x - 1}{x}.$$

Example 2. *Simplify* $\dfrac{x^4 - 27x}{x^2 - 9} \div \dfrac{x^2 + 3x + 9}{x + 3}.$

Factorising the expression is equal to:

$$\frac{x(x^3 - 27)}{x^2 - 9} \div \frac{x^2 + 3x + 9}{x + 3}$$

$$= \frac{x(x - 3)(x^2 + 3x + 9)}{(x + 3)(x - 3)} \times \frac{x + 3}{x^2 + 3x + 9}$$

$$= x \quad \text{(on cancelling factors).}$$

Exercise 31.

Simplify the following fractions:

1. $\dfrac{xy}{x^2 - 1} \times \dfrac{x^2 - x}{y}.$

2. $xy \div \dfrac{x}{y + 1}.$

3. $\dfrac{a^2 - b^2}{a^2 + 2ab + b^2} \times \dfrac{ab + b^2}{a^2 - ab}.$

4. $\dfrac{a^2 - 49}{a^2 - 9} \div \dfrac{a + 7}{a + 3}.$

5. $\dfrac{2a^3 - 6a^2b}{2a^3 - 2ab^2} \times \dfrac{a^2 + 3ab + 2b^2}{a - 3b}.$

6. $\dfrac{2x^2 + x - 1}{x - 2} \div \dfrac{2x^2 - x}{x^2 - 5x + 6}.$

7. $\dfrac{2a^2 - 8}{a + 2} \div 2a - 4.$

8. $\dfrac{6x^2 - x - 2}{8x + 4} \times \dfrac{2}{3x^2 - 5x + 2}.$

9. Simplify $\dfrac{2b}{b^2 - 1} \times \dfrac{b^2 - 3b - 4}{b^2}$ and find its value when $b = 5$.

10. Simplify $\dfrac{x^2 + 6x + 9}{2x - 2} \times \dfrac{x^2 - 1}{x^2 + 4x + 3}$ and find its value when $x = 1·5$.

---

102. Addition and subtraction.

The fundamental principles were examined in § 23. We now proceed to more difficult examples.

Example 1. *Simplify* $\dfrac{a}{a - b} - \dfrac{a^2}{a^2 - b^2}.$

Factorising, the expression becomes

$$\frac{a}{a - b} - \frac{a^2}{(a + b)(a - b)}.$$

The least common denominator is $(a + b)(a - b)$.
∴ fraction

$$= \frac{a(a + b) - a^2}{(a + b)(a - b)}$$
$$= \frac{a^2 + ab - a^2}{(a + b)(a - b)}$$
$$= \frac{ab}{(a + b)(a - b)}.$$

Example 2. *Simplify* $\dfrac{3}{a - b} - \dfrac{2a + b}{a^2 - b^2}.$

Factorising, the expression becomes

$$\frac{3}{a-b} - \frac{2a+b}{(a+b)(a-b)}$$

$$= \frac{3(a+b) - (2a+b)}{(a+b)(a-b)} = \frac{3a+3b-2a-b}{(a+b)(a-b)}$$

$$= \frac{a+2b}{(a+b)(a-b)}.$$

This example is intended to remind the student of the note at the end of § 28 relative to a minus sign before a fraction. The numerator, as $(2a+b)$ in the above example, should always be placed in a bracket on addition of the fraction, and the bracket removed afterwards.

Example 3. *Simplify* $\dfrac{x-1}{x^2-x-2} - \dfrac{x+2}{x^2+4x+3}$.

Factorising, the fraction becomes

$$= \frac{x-1}{(x-2)(x+1)} - \frac{x+2}{(x+3)(x+1)}$$

$$= \frac{(x-1)(x+3) - (x+2)(x-2)}{(x+1)(x-2)(x+3)}$$

$$= \frac{(x^2+2x-3) - (x^2-4)}{(x+1)(x-2)(x+3)}$$

$$= \frac{x^2+2x-3-x^2+4}{(x+1)(x-2)(x+3)}$$

$$= \frac{2x+1}{(x+1)(x-2)(x+3)}.$$

Example 4. *Simplify* $\dfrac{x}{x-\dfrac{1}{x}}$.

$$\frac{x}{x-\dfrac{1}{x}} = \frac{x}{\dfrac{x^2-1}{x}} = x \div \frac{x^2-1}{x}$$

$$= x \times \frac{x}{x^2-1} = \frac{x^2}{x^2-1}.$$

Example 5. *Convert the formula* $\dfrac{1}{R} = \dfrac{1}{R_1} - \dfrac{1}{R_2}$ *into one in which the subject is R.*

$$\frac{1}{R} = \frac{1}{R_1} - \frac{1}{R_2}$$

i.e.,
$$\frac{1}{R} = \frac{R_2 - R_1}{R_1 R_2}.$$

$$\therefore \quad R = \frac{R_1 R_2}{R_2 - R_1} \quad \left(\text{just as, if } \frac{1}{R} = \frac{1}{4}, \; \therefore \; R = 4 \right).$$

Note.—The student should not make the mistake of inverting at the outset. The fractions on the right side must be added first.

Exercise 32.

Simplify the following:

1. $\dfrac{1}{x} + \dfrac{1}{x - y}.$

2. $\dfrac{2}{x - 1} - \dfrac{3}{x - 2}.$

3. $\dfrac{2}{3(a - b)} + \dfrac{3}{2(a + b)}.$

4. $1 - \dfrac{1}{x} + \dfrac{1}{x + 2}.$

5. $3 - \dfrac{2}{x - 3} + \dfrac{1}{(x - 3)^2}.$

6. $\dfrac{2x}{3(x + 1)} - \dfrac{3x}{2(x - 2)}.$

7. $\dfrac{3}{1 - a} + \dfrac{4}{(1 - a)^2}.$

8. $\dfrac{x + 3}{x - 3} - \dfrac{x - 3}{x + 3}.$

9. $\dfrac{1}{x} - \dfrac{1}{x + 2y} - 1.$

10. $\dfrac{b}{3a - b} - \dfrac{a}{3b} + 1.$

11. $\dfrac{1}{1 - at} - \dfrac{1}{1 - bt}.$

12. $\dfrac{1}{a + bt} - \dfrac{1}{x + yt}.$

13. $\dfrac{1}{x + y} + \dfrac{1}{x - y} - \dfrac{1}{x^2 - y^2}.$

14. $\dfrac{x}{x^2 - y^2} - \dfrac{y}{(x - y)^2} - \dfrac{1}{x + y}.$

Transform the following formulae to those in which the subject is R:

15. $\dfrac{1}{R} = \dfrac{1}{p} + \dfrac{1}{q}.$

16. $\dfrac{1}{R} = \dfrac{1}{p} - \dfrac{1}{q}.$

17. $\dfrac{1}{R} = \dfrac{2}{R_1} + \dfrac{3}{R_2}.$

18. $\dfrac{1}{R} = \dfrac{3}{R_1} - \dfrac{2}{R_2}.$

19. If $\dfrac{1}{R} = \dfrac{1}{r+s} - \dfrac{1}{r-s}$ find R in terms of the other letters.

20. If $\dfrac{1}{P} = \dfrac{2}{p-q} + \dfrac{3}{p+q}$ find the value of $5P$.

21. Simplify $P + Q - \dfrac{P^2 + Q^2}{P+Q}$.

22. Simplify $\dfrac{P^2 + Q^2}{P-Q} - P$.

23. If $\dfrac{1}{R} = \dfrac{1}{R_1} + \dfrac{1}{R_2} + \dfrac{1}{R_3}$ find R when $R_1 = 8 \cdot 6$, $R_2 = 4 \cdot 3$, and $R_3 = 3$.

24. Given $a(P - \tfrac{1}{2}Q) = b(Q - \tfrac{1}{2}P)$ rearrange the terms so as to express P in terms of the other quantities.

25. Given $I = \dfrac{nE}{R + nr}$ find n in terms of the other quantities. Find n when $I = 2$, $E = 1 \cdot 8$, $R = 2 \cdot 4$, $r = 0 \cdot 5$.

---

103. Simple equations involving algebraical fractions.

The following examples will illustrate the methods employed.

Example I. *Solve the equation:*

$$\frac{3}{x-2} = \frac{5}{x-1}.$$

The least common denominator is $(x-2)(x-1)$.
We therefore multiply both sides by this.

Then $\qquad \dfrac{3(x-2)(x-1)}{x-2} = \dfrac{5(x-2)(x-1)}{x-1}$.

Cancelling $\qquad 3(x-1) = 5(x-2)$
or $\qquad 3x - 3 = 5x - 10$.

$\therefore \qquad\qquad 2x = 7$

and $\qquad\qquad x = 3 \cdot 5$.

Example 2. *Solve for n the equation:*

$$\frac{1}{n-2} + \frac{1}{n-3} = \frac{2}{n}.$$

Multiply throughout by the least common denominator $n(n-2)(n-3)$.

Then $n(n-3) + n(n-2) = 2(n-2)(n-3)$.

$\therefore \qquad n^2 - 3n + n^2 - 2n = 2(n^2 - 5n + 6)$

or $\qquad\qquad 2n^2 - 5n = 2n^2 - 10n + 12$.

The terms involving $2n^2$ disappear.

$\therefore \qquad\qquad\qquad 5n = 12$

and $\qquad\qquad\qquad n = \dfrac{12}{5}$.

Exercise 33.

Solve the following equations:

1. $\dfrac{18}{2x} - 1 = 2$.

2. $\dfrac{4x}{2x-1} = 5$.

3. $\dfrac{x-1}{x-2} = 3$.

4. $\dfrac{1-x}{x+1} = 4$.

5. $0 \cdot 8 = \dfrac{1 \cdot 5n}{4 + 1 \cdot 4n}$.

6. $\dfrac{5}{2x+5} = \dfrac{4}{x+5}$.

7. $\dfrac{3p+23}{3p+12} = \dfrac{4}{3}$.

8. $\dfrac{\dfrac{1}{x} - 3}{\dfrac{1}{x} - 2} = \dfrac{1}{4}$.

9. $\dfrac{3}{x-2} - \dfrac{5}{x+1} = \dfrac{5}{(x-2)(x+1)}$.

10. $\dfrac{3}{x} - \dfrac{2}{x-2} = 0$.

11. In the formula $\dfrac{1}{f} = \dfrac{1}{v} - \dfrac{1}{u}$ find v when $f = 8$, $u = 2$.

12. For what value of n is $\dfrac{5n-20}{4n+6}$ equal to $\tfrac{1}{3}$?

13. If $C = \dfrac{V}{R}$ find R when $C = 7 \cdot 5$, $V = 60$.

14. If $r = \dfrac{R(E-V)}{V}$ find V in terms of r, R, E.

CHAPTER XIII

GRAPHS OF QUADRATIC FUNCTIONS

104. Constants and variables.

In the formulae which have been considered in earlier chapters it will be seen that the letters employed represent two different kinds of numbers.

(1) Some represent **constant numbers**—*i.e.*, numbers which remain unchanged in the varying cases to which the formula applies.

(2) Others represent **variable quantities**.

Example 1. Consider the formula for the circumference of a circle, viz.

$$C = 2\pi r,$$

where $C =$ the length of the circumference,

$r = \quad ,, \quad ,, \quad$ radius.

Two of these four symbols represent constants.

(1) The number 2 (which is fixed for any circle),

(2) The number π (which has always the same value in any formula in which it is used). It represents the **constant ratio** of the circumference of any circle to its diameter, or the area of a circle to the square of its radius.

Two of the letters in the formula are variables, they are different in value for different circles, the variation of C depending on changes in r.

Example 2. In § 68 it was shown that the formula, or equation connecting profit and customers, was

$$y = 0 \cdot 05x - 11 \cdot 5.$$

It was pointed out that while the profit represented by y depended on the varying number of customers x, the numbers $0 \cdot 05$ and $11 \cdot 5$ remained constant, representing respectively the average amount paid by each customer, and the fixed charges.

Example 3. This last example was a special case of the equation to a straight line, which in § 74 was shown to be represented in general by the equation:

$$y = mx + b.$$

For a particular straight line, m is constant, representing the gradient of the line, while b is the fixed distance intercepted on the y axis. But x and y vary for different points on the line. They represent the co-ordinates of **any** point.

105. Dependent and independent variables.

The variables are seen to be of two kinds.

Considering the case of the circumference of the circle, the length of this **depends** on the length of the radius.

In the second example the profit **depends** on the number of customers.

A variable which thus depends on another variable for its value is called a **dependent variable**.

The other variable upon which the first one depends is called the **independent variable**.

As another example, if a train is moving with uniform —*i.e.*, constant—speed, the **distance travelled depends on the time**.

Thus the speed is a **constant**,

time is an **independent variable**,
distance is a **dependent variable**.

Again the cost of a quantity of tea depends on the weight bought, the price per kilo being constant.

Thus weight is the independent variable.
cost ,, dependent variable.

Graphs. In plotting graphs which show how one quantity varies as another varies, the independent variable is always measured on the x axis and the dependent on the y axis.

106. Functions.

When two quantities are connected as shown above, **the dependent variable is said to be a function of the independent variable.** Thus:

The **circumference** of a circle is a function of its **radius**.

The **area** of a square is a function of its side.

The **distance** travelled by a car moving uniformly is a function of the **time**.

If a spring is stretched by a force, the **extension** of the spring is a function of the **force**.

Generally if a quantity denoted by y depends for its value on another quantity x then **y is a function of x.**

Thus in each of the following examples

$$y = 2x$$
$$y = x^2$$
$$y = \frac{5}{x} - 3$$
$$y = 3(x-1)(x-3)$$

the value of y depends on the value of x. If any particular value be given to x, a corresponding value of y can be calculated. In all such cases y is a function of x.

The idea of a function is probably the most important in modern Mathematics. From the above examples the following definition can be deduced:

Function. If a quantity y is related to a quantity x so that for every value which might be assigned to x there is a corresponding value of y, then y is a function of x.

Thus for every length that may be chosen for the radius of a circle, there is always a corresponding length of the circumference.

If a man works at a fixed rate per hour, then for any number of hours that he works, there is always a corresponding amount of pay. His pay is a function of the time he works; the rate per hour is a constant.

Again if $y = 2x^3$, then for every value which we may choose to give to x, there is a corresponding value of y. Consequently, y is a function of x.

107. Graph of a function.

If y is a function of x, and since, by the above definition, for every value assigned to x there is always a corresponding value of y, then these pairs of values of x and y can be plotted, and the assemblage of points so plotted will be the graph of the function.

Thus every function has a distinctive graph, which will be a regular curve or a straight line, and by which it can be identified.

If the function is of the first degree, and does not involve any algebraical fraction, then, as we have already seen, the graph will be a straight line. For this reason a function of the first degree, of which the general form is $y = mx + b$, is called a linear function.

If, however, the function involves a higher power of x, such as x^2, x^3, etc., or involves an algebraic fraction such as $y = \dfrac{1}{x}$, the graph will be a regular curve, the shape of which will differ with the nature of the function.

108. Graph of a function of second degree.

The simplest form of a function of the second degree is that which is expressed by $y = x^2$. This is called a quadratic function, from the Latin quadratus (squared). The area of a circle, $A = \pi r^2$, is a special form of this.

To plot the curve of $y = x^2$. We first assign values to x, calculate the corresponding values of x^2, or y, and tabulate them as follows:

x	-3	$-2 \cdot 5$	-2	$-1 \cdot 5$	-1	0	1	$1 \cdot 5$	2	$2 \cdot 25$	3
y	9	$6 \cdot 25$	4	$2 \cdot 25$	1	0	1	$2 \cdot 25$	4	$6 \cdot 25$	9

As wide a range of values is taken as the size of the paper will allow to be plotted. Since the values of y increase more rapidly than those of x, more room is needed on the y axis, but as the square of a number is always positive, no negative values of y are necessary. The x axis is therefore drawn near the bottom of the paper, as shown in Fig. 31.

Selection of scales on axes. The scale chosen must depend on the number of values we wish to include within the limits of the paper. The same scales need not be chosen on the two axes, and in this particular curve it will be better to take a smaller scale for values of y, because their values increase more rapidly than those of x.

It must be remembered, however, that the true shape

of the curve will be shown only when the same units are taken on both axes.

The symmetry of the values of y, as shown in the table, suggests that the y axis should be in the middle of the paper.

The points when plotted appear to lie on a smooth curve as shown in Fig. 31.

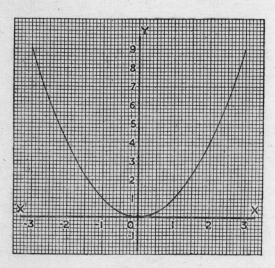

Fig. 31.

The curve is called a **parabola**. When inverted it is the curve described by a projectile such as a shot from a gun, or a rocket when fired into the air.

109. Some properties of the curve of $y = x^2$.

The following points should be noted about this curve:

(1) **The symmetry of the curve.** Positive and negative values of x produce equal values for y. If therefore the curve were folded about the y axis, the two parts of the curve, point by point, would coincide.

The curve is therefore said to be **symmetrical about** OY, and OY is called the **axis of symmetry**.

(2) The **minimum value** of the curve is 0 at the origin. The curve is said to have a turning-point at the origin.

(3) The **slope** of the curve is not constant, as with a straight line, but increases from point to point as x increases. The gradient is clearly a function of x, since its value depends on the value of x.

(4) The curve may be used to read off the square of any number within the range of plotted values, and also, conversely, to determine square roots. Thus to find $\sqrt{3}$, take the point on the curve corresponding to 3 on the y axis. It is then seen that there are two points on the x axis which correspond to this, the values of x (i.e., $\sqrt{3}$) being $+1\cdot73$ and $-1\cdot73$ (see § 43).

110. The curve of $y = -x^2$.

All values of y for this curve are equal numerically to the corresponding values of y in $y = x^2$, but are negative. The shape of the curve will be the same, but inverted, as in Fig. 32.

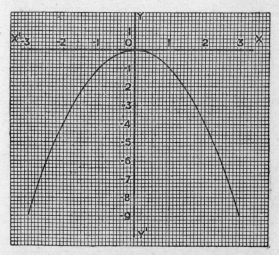

FIG. 32.

All the values of y being negative, the x axis is drawn toward the top of the paper.

The curve has a **maximum** point—viz. zero—at the origin. From this point the curve shows the path of a bomb dropped from an aeroplane at 0—ignoring air resistance.

III. The curves of $y = ax^2$.

The curves represented by $y = ax^2$, where a is any number, are all parabolas differing from $y = x^2$ only in having different slopes.

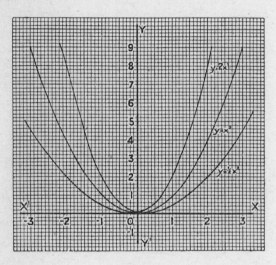

Fig. 33.

Considering $y = 2x^2$ as an example the table of values for plotting the curve are found in the same way as for $y = x^2$.

This curve, as well as that of $y = \frac{1}{2}x^2$, are shown in Fig. 33, and contrasted with $y = x^2$.

If a be negative, we get a corresponding set of curves similar to $y = -x^2$, as in Fig. 32.

112. The curves of $y = x^2 \pm a$, where a is any number.

The curve of $y = x^2 + 2$ is evidently related to that of $y = x^2$ in the same way that $y = x + 2$ is connected with $y = x$ (see § 74). Each ordinate of $x^2 + 2$ is greater by 2 than the corresponding ordinate of x^2.

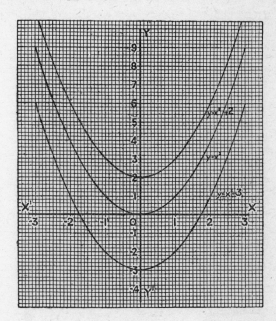

Fig. 34.

The curve of $y = x^2 + 2$ is therefore the same as $y = x^2$ raised two units on the y axis. It appears as shown in Fig. 34. Similarly, the curve $y = x^2 - 3$ is the curve of $y = x^2$, but every point is three units lower than $y = x^2$ for corresponding values of x.

Generally the curves represented by the equation $y = x^2 \pm a$ are a set of curves similar to $y = x^2$ and raised or lowered by an amount equal to $\pm a$, according to the sign of a.

Similarly, the curve of $y = -x^2$ gives rise to a set of curves included in the equation $y = -x^2 \pm a$.

113. Change of axis.

On examination of Fig. 34, it will be seen that the curve of $y = x^2 + 2$ differs from the curve of $y = x^2$ only in the position of the x axis. For the curve of $y = x^2 + 2$ the

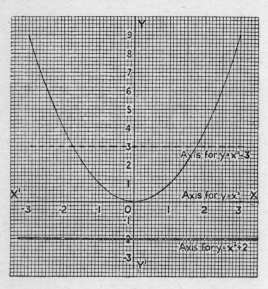

FIG. 35.

x axis is two units below the lowest point of the curve. Consequently the curve of $y = x^2$ can be changed into that of $y = x^2 + 2$ by drawing a new x axis two units below the original, as shown in Fig. 35, and re-numbering the y axis.

Similarly, $y = x^2$ becomes $y = x^2 - 3$ by drawing a new x axis three units above the original, as shown, and again re-numbering the units on the y axis.

114. Curve of $y = (x - 1)^2$.

The table below shows the values of (1) $y = (x - 1)^2$
and (2) $y = x^2$, for corresponding values of x.

x . .	-3	-2	-1	0	1	2	3	4
x^2 . .	9	4	1	0	1	4	9	16
$(x - 1)^2$.	16	9	4	1	0	1	4	9

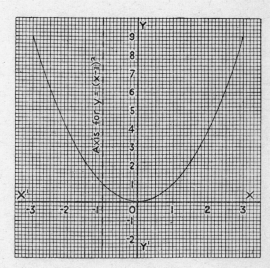

Fig. 36.

A comparison of the sets of values of two functions shows
that both have the **same sequence of values**, but those of
$(x - 1)^2$ are those of x^2 **moved one place to the right** in
the table for consecutive values of x.

Consequently the curve of $(x - 1)^2$ must be the same as
that of x^2, but moved one unit to the right.

Therefore, as in the previous paragraph, but moving
the y axis instead of the x axis, the curve of x^2 will be
changed into the curve of $(x - 1)^2$ by moving the **y axis**

one unit to the left and re-numbering the x axis (see Fig. 36).

Similarly if the y axis be moved two units to the left, we get the curve of $(x - 2)^2$. Or if the y axis be moved one unit to the **right** we should have the curve of $y = (x + 1)^2$, and so for similar functions.

115. Graph of $y = (x - 1)^2 - 4$.

By combining the operations illustrated in the preceding two paragraphs—*i.e.*, by moving both axes—the curve of

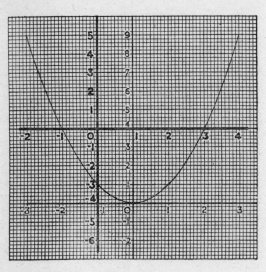

Fig. 37.

$y = x^2$ can be changed into the curve of such a function as $y = (x - 1)^2 - 4$. This can be done in two steps:

(1) By moving the y axis one unit to the left we get the curve of $y = (x - 1)^2$, as in § 114.

(2) By now moving the x axis 4 units up the y axis the curve of $y = (x - 1)^2$ becomes $y = (x - 1)^2 - 4$.

With these two new axes, both re-numbered, the curve is as shown in Fig. 37.

In this diagram the new axes have been drawn and the new numbers on the axes marked in larger figures.

The expression may be simplified, since

$$(x - 1)^2 - 4 = x^2 - 2x + 1 - 4$$
$$= x^2 - 2x - 3.$$

The curve in Fig. 37 is therefore the graph of

$$y = x^2 - 2x - 3.$$

We shall see later (see § 121) that every quadratic function of x can be reduced to such a form as this, or the simpler forms of §§ 113 and 114. We may therefore draw the conclusion that:

The graph of every quadratic function of x is a modified form of $y = x^2$ and is therefore a parabola.

116. The graph $y = x^2 - 2x - 3$.

We have seen in the preceding paragraph that this graph can be obtained from the curve of $y = x^2$ by changing axes. In practice, however, the method given below will be found more convenient in most cases and usually more accurate in practical application. A table of values is constructed as follows:

x	-3	-2	-1	0	1	2	3	4
x^2	9	4	1	0	1	4	9	16
$-2x$	$+6$	$+4$	$+2$	0	-2	-4	-6	-8
-3	-3	-3	-3	-3	-3	-3	-3	-3
y	12	5	0	-3	-4	-3	0	5

Note.—Be careful not to add the value of x in each column. It is well to draw a thick line under the values as shown, to remind one when adding.

From the values in this table the student should plot the curve and compare the figure obtained with that in Fig. 37, arrived at by change of axes. It will be seen that the minimum value of $x^2 - 2x - 3$ is -4, when $x = 1$, and the line $x = 1$, perpendicular to the x axis, is an axis of symmetry.

117. Solution of the equation $x^2 - 2x - 3 = 0$ from the graph.

At the points where the curve of $x^2 - 2x - 3$, in Fig. 37, cuts the axis of x, the value of y is zero—*i.e.*:

$$x^2 - 2x - 3 = 0.$$

The values of x at these two points are:

$$x = 3 \text{ and } x = -1.$$

We have thus found two values of x which satisfy the equation $x^2 - 2x - 3 = 0$. The equation has therefore been solved by means of the graph.

118. Graph of $y = 2x^2 - 3x - 5$.

This is a slightly more difficult graph to plot. The table of values is as follows:

x .	-3	-2	-1	0	$\frac{1}{2}$	1	2	3	4	5
$2x^2$.	18	8	2	0	$\frac{1}{2}$	2	8	18	32	50
$-3x$.	9	6	3	0	$-1\frac{1}{2}$	-3	-6	-9	-12	-15
-5 .	-5	-5	-5	-5	-5	-5	-5	-5	-5	-5
y .	22	9	0	-5	-6	-6	-3	4	15	30

In this example a wider range of values is shown. As the values of y increase rapidly smaller units are taken on the two axes. The graph is as shown in Fig. 38.

The lowest point, N, giving the **minimum** value of the expression corresponds to $x = \frac{3}{4}$. From the table of values this point is seen to be half-way between $x = \frac{1}{2}$ and $x = 1$, since these values of x give the same values of y—viz., -6. The minimum value, represented by MN, is $-6\frac{1}{8}$. The ordinate through M, of which MN is part, is the axis of symmetry of the curve.

The solution of the equation:

$$2x^2 - 3x - 5 = 0,$$

will be given by the values of x, where $y = 0$—*i.e.*, at A and B, where the curve cuts the axes. These points give:

$$x = -1, \text{ or } 2\cdot5.$$

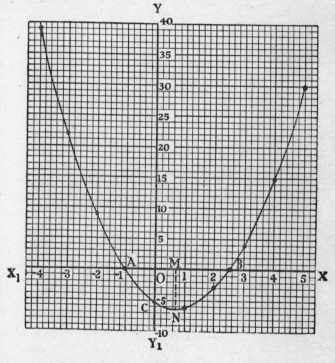

FIG. 38.

119. Graph of $y = 12 - x - x^2$.

In this example the coefficient of x^2 is negative. Consequently the curve will take the form of $y = -x^2$, as shown in Fig. 32—*i.e.*, it will be an inverted parabola. The table of values is as follows:

x .	-5	-4	-3	-2	-1	0	1	2	3	4
12 .	12	12	12	12	12	12	12	12	12	12
$-x$.	5	4	3	2	1	0	-1	-2	-3	-4
$-x^2$	-25	-16	-9	-4	-1	0	-1	-4	-9	-16
y .	-8	0	6	10	12	12	10	6	0	-8

The values of y show the symmetry of the curve. The highest point, giving the **maximum** value, is seen to be

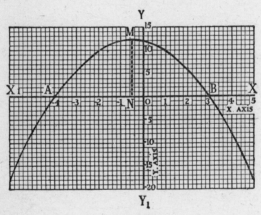

Fig. 39.

half-way between $x = -1$, and $x = 0$—i.e., where $x = -\frac{1}{2}$. The curve is as shown in Fig. 39. The maximum value is at M, where the value of y is $11\frac{1}{4}$.

The ordinate from M—i.e., MN—is the axis of symmetry for the curve.

We can obtain from the curve the solution of the equation:

$$12 - x - x^2 = 0 \text{—}i.e.\ x^2 + x - 12 = 0.$$

This will be given by the values of x at A and B, where the curve cuts the x axis.   The solutions are:

$$x = 3, \, x = -4.$$

Exercise 34.

1. Draw the curve of $y = x^2$ between the values $x = +3$ and $x = -3$, taking the units as large as possible.   From the curve write down the values of:

    (1) $2 \cdot 3^2$.   (2) $\sqrt{7}$.   (3) $\sqrt{3 \cdot 5}$.

2. Draw the curve of $y = \frac{1}{4}x^2$ between $x = +4$ and $x = -4$.   Find from the curve the values of x such that:

    (1) $\frac{1}{4}x^2 = 0 \cdot 8$.   (2) $\frac{1}{4}x^2 = 2$.   (3) $x^2 = 12$.

3. Draw the curve of $-\frac{1}{2}x^2$ between $x = +4$ and $x = -4$.   From the curve find the values of:

    (1) $-\frac{1}{2}x^2 = -3$.   (2) $-x^2 = -5$.   (3) $x^2 = 2$.

4. Draw the curve of $y = x^2$ and by change of axes obtain the curve of $y = x^2 + 3$.   From the curve find the values of x such that :

    (1) $x^2 + 3 = 5$.   (2) $x^2 + 3 = 9$.

5. Draw the curve of $y = x^2$ and by change of axes obtain the curves of:

    (a) $y = (x - 2)^2$.   (b) $y = (x + 3)^2$.

From the curves find the values of x such that:

    (1) $(x - 2)^2 = 3$.   (2) $(x + 3)^2 = 1$.
    (3) $(x - 2)^2 - 5 = 0$.

6. Draw the curve of $y = (x + 2)^2 - 2$ by means of the curve of $y = x^2$ and changes of axes.   Use the curve to find the values of x when:

    (1) $(x + 2)^2 - 2 = 0$.   (2) $(x + 2)^2 - 1 = 0$.
    (3) $(x + 2)^2 = 8$.

7. Draw the curve of $y = x^2 - 6x + 5$.   Find the least value of this function and the corresponding value of x.   Use the curve to find the values of x when:

    (1) $x^2 - 6x + 5 = 0$.   (2) $x^2 - 6x + 5 = 6$.

8. Draw the curve of $y = x^2 - 4x + 2$. Find the minimum value of the function and the corresponding value of x. Use the curve to solve the following equations:

(1) $x^2 - 4x - 2 = 0$.   (2) $x^2 - 4x - 2 = 3$.

9. Draw the curve of $y = 2x^2 - 5x + 2$. Find the minimum value of the function and the corresponding value of x. From the curve solve the equations:

(1) $2x^2 - 5x + 2 = 0$.   (2) $2x^2 - 5x - 1 = 0$.

10. Draw the curve of $y = 2 - x - x^2$. Find the maximum value of the function and the corresponding value of x.

CHAPTER XIV

QUADRATIC EQUATIONS

120. Algebraical solution.

PLOTTING the graph of a quadratic function led logically to the solution of a quadratic equation—*i.e.*, an equation of the second degree. The solution by this method is useful and illuminating, but as a method of solving a quadratic equation it is cumbersome, and the accuracy obtainable is limited. We must therefore find an algebraical solution which is certain, universally applicable and capable of any required degree of accuracy.

A quadratic equation was solved for the first time when, in § 109, from the curve of $y = x^2$ it was found that if $x^2 = 3$, the corresponding values of x were $+ 1\cdot73$ and $- 1\cdot73$. The reasoning may be stated thus:

$$x^2 = 3.$$
$$\therefore \qquad x = \pm \sqrt{3}$$
$$\text{and} \qquad x = + 1\cdot73 \text{ or } - 1\cdot73.$$

This is the simplest form of a quadratic equation. It involves the operation of finding a square root; hence the term " root " as applied to the solution of an equation.

In § 115 an important step forward was made. The points where the curve of $(x - 1)^2 - 4$ cuts the x axis, and the function was therefore equal to zero, were found, and the corresponding values of x noted—viz., 3 and $- 1$.

This means that for these values of x, the expression

$$(x - 1)^2 - 4 = 0.$$

They are therefore the roots of this equation.
Let the equation be written in the form:

$$(x - 1)^2 = 4 \quad . \quad . \quad . \quad . \quad \text{(A)}$$

Then algebraically it is of the same form as the equation

above, $x^2 = 3$. We can proceed with the algebraical solution on the same lines.

Taking the square roots of each side:

$$x - 1 = \pm 2.$$
$$\therefore \qquad x - 1 = + 2 \text{ or } x - 1 = - 2,$$
whence $\qquad x = 3 \text{ or } x = - 1.$

The form marked (A) is the form to which, ultimately, all quadratic equations are reduced; our object always is to reach this form.

121. The method of solution of any quadratic.

It was shown that the expression $(x - 1)^2 - 4$ simplified to $x^2 - 2x - 3$.

$\therefore$ the equation which was solved could have been written in the form:

$$x^2 - 2x - 3 = 0 \quad . \quad . \quad . \quad . \quad (1)$$

If we start with this equation and wish to solve it, we need to get back to the form:

$$(x - 1)^2 - 4 = 0 \quad . \quad . \quad . \quad . \quad (2)$$

This is the converse operation of changing from (2) to (1), *i.e.* of obtaining the complete square (2) when we are given (1).

Two preliminary steps are necessary.

(1) Remove the constant to the right side, as it does not help in finding the square.

(2) Divide throughout by the coefficient of x^2, if this is not unity.

Finally we arrive at the form:

$$x^2 - 2x = 3.$$

It is now necessary to add to the left side such a number ae will produce a complete square. Remembering the work in § 92, we get the following rule:

Add to each side the square of half the coefficient of x.

Then the above becomes

$$x^2 - 2x + (1)^2 = 3 + 1.$$
$$\therefore \qquad (x - 1)^2 = 4,$$

and we proceed as before.

We will now apply the above method to solve the equation:

$$x^2 - 6x + 5 = 0.$$
Then $\qquad x^2 - 6x = -5.$

Add *to each side* $(\frac{6}{2})^2$—*i.e.*, 3^2.
Then $\qquad x^2 - 6x + (3)^2 = -5 + 9.$
$\therefore \qquad (x - 3)^2 = 4.$

Thus we reach the desired form. Proceeding as before:

$$x - 3 = \pm 2.$$
$$\therefore \qquad x = 3 \pm 2.$$
Then $\qquad x = 3 + 2 = 5$
or $\qquad x = 3 - 2 = 1.$

$\therefore$ the solution is $x = 5$ or $x = 1$.

122. Solution of $2x^2 + 5x - 3 = 0$.

Applying the preliminary steps (1) and (2) of the previous paragraph, we get in succession:

$$2x^2 + 5x = 3$$
and $\qquad x^2 + \frac{5}{2}x = \frac{3}{2}.$

Half the coefficient of x is $\frac{5}{4}$.
Adding $(\frac{5}{4})^2$ we get:

$$x^2 + \frac{5}{2}x + (\frac{5}{4})^2 = \frac{3}{2} + (\frac{5}{4})^2.$$
$$\therefore \qquad (x + \frac{5}{4})^2 = \frac{3}{2} + \frac{25}{16} = \frac{49}{16}.$$

Taking square roots of both sides:

$$x + \frac{5}{4} = \pm \frac{7}{4}.$$
$$\therefore \qquad x = -\frac{5}{4} \pm \frac{7}{4}$$
and $\qquad x = -\frac{5}{4} + \frac{7}{4} = \frac{2}{4}$
or $\qquad x = -\frac{5}{4} - \frac{7}{4} = \qquad \frac{12}{4}.$

$\therefore \qquad x = \frac{1}{2}$ or $x = -3$.

123. Worked examples.

Example 1. *Solve the equation $x^2 - x - 1 = 0$.*

Transposing

$$x^2 - x = 1.$$

Adding $(\frac{1}{2})^2$

$$x^2 - x + (\tfrac{1}{2})^2 = 1 + \tfrac{1}{4}$$

or

$$(x - \tfrac{1}{2})^2 = \tfrac{5}{4}.$$

$$\therefore \qquad x - \tfrac{1}{2} = \pm \frac{\sqrt{5}}{2}$$

and

$$x = \tfrac{1}{2} \pm \frac{\sqrt{5}}{2}.$$

$$\therefore \qquad x = \tfrac{1}{2} + \frac{\sqrt{5}}{2} = \frac{1 + 2 \cdot 236}{2} = 1 \cdot 618$$

or

$$x = \tfrac{1}{2} - \frac{\sqrt{5}}{2} = \frac{1 - 2 \cdot 236}{2} = -0 \cdot 618.$$

$$\therefore \qquad x = 1 \cdot 618 \text{ or } x = -0 \cdot 618 \text{ (both approx.)}.$$

Example 2. *Solve the equation $3x^2 - 5x + 1 = 0$.*

Applying preliminary steps, § 121:

$$3x^2 - 5x = -1$$
$$x^2 - \tfrac{5}{3}x = -\tfrac{1}{3}.$$

Adding $(\frac{5}{6})^2$:

$$x^2 - \tfrac{5}{3}x + (\tfrac{5}{6})^2 = -\tfrac{1}{3} + \tfrac{25}{36}.$$
$$\therefore \qquad (x - \tfrac{5}{6})^2 = \tfrac{13}{36}.$$

Taking square roots

$$x - \tfrac{5}{6} = \frac{\pm 3 \cdot 606 \text{ (approx.)}}{6}.$$

$$\therefore \qquad x = \frac{5 \pm 3 \cdot 606}{6}.$$

$$\therefore \qquad x = \frac{5 + 3 \cdot 606}{6} = \frac{8 \cdot 606}{6} = 1 \cdot 434$$

or

$$x = \frac{5 - 3 \cdot 606}{6} = \frac{1 \cdot 394}{6} = 0 \cdot 232.$$

$\therefore$ solution is $x = 1 \cdot 434$ or $x = 0 \cdot 232$ (approx.).

Example 3. *Solve the equation:*

$$\frac{1}{x - 1} - \frac{1}{x + 2} = \frac{1}{16}.$$

First we clear fractions by multiplying throughout by the least common denominator—viz., $16(x-1)(x+2)$.

Then $16(x+2) - 16(x-1) = (x-1)(x+2)$.

∴ $16x + 32 - 16x + 16 = x^2 + x - 2$.

Adding and transposing:

$$-x^2 - x = -48 - 2$$

or $$x^2 + x = 50.$$

Adding $(\frac{1}{2})^2$:

$$x^2 + x + (\tfrac{1}{2})^2 = 50 + \tfrac{1}{4} = \tfrac{201}{4}.$$

∴ $$(x + \tfrac{1}{2})^2 = \pm \frac{\sqrt{201}}{2} = \frac{\pm\ 14 \cdot 177 \text{ (approx.)}}{2}.$$

∴ $$x = -\tfrac{1}{2} \pm \frac{14 \cdot 177}{2}$$

and $$x = -\tfrac{1}{2} + \frac{14 \cdot 177}{2} = \frac{13 \cdot 177}{2} = 6 \cdot 588.$$

or $$x = -\tfrac{1}{2} - \frac{14 \cdot 177}{2} = -\frac{15 \cdot 177}{2} = -7 \cdot 588.$$

∴ the solution is $x = 6 \cdot 588$ or $x = -7 \cdot 588$.

Exercise 35.

Solve the following equations:

1. $3x^2 = 12$.
2. $4x^2 - 1 = 0$.
3. $\dfrac{x^2}{16} = 9$.
4. $(x+1)^2 - 4 = 0$.
5. $(x-3)^2 - 25 = 0$.
6. $(x+5)^2 = 36$.
7. $(x + \tfrac{1}{4})^2 = 1$.
8. $\dfrac{(x-9)^2}{15} = 4$.
9. $x^2 - 10x + 16 = 0$.
10. $x^2 + x - 12 = 0$.
11. $x^2 - 2x - 15 = 0$.
12. $x^2 + 3x - 28 = 0$.
13. $x(x-4) = 32$.
14. $2x^2 - 7x + 6 = 0$.
15. $2x^2 - 3x - 5 = 0$.
16. $3x^2 = 7x + 9$.
17. $3x^2 + 1 = 5x$.
18. $\dfrac{6}{x} + \dfrac{3}{x+2} = 2$.
19. $3x - \dfrac{5}{x} = 14$.
20. $x - 2 = \dfrac{2}{x}$.
21. $\dfrac{x-9}{3} = \dfrac{x+5}{x}$.
22. $\dfrac{1}{x-1} - \dfrac{1}{x+2} = \dfrac{1}{16}$.

124. Solution of quadratic equations by factorisation.

There is another method of solving quadratics; it is as follows:

We first note that if n be any finite number

$$n \times 0 = 0,$$

i.e., the product of any finite number and zero is always zero.

Conversely, if the product of two factors is zero, then either of the factors may be zero.

For example, if $a \times b = 0$
then this is true if either $a = 0$, or $b = 0$.

Similarly if

$$(x - 1)(x - 3) = 0 \quad . \quad . \quad . \quad \text{(A)}$$

it follows from the above that this equation is satisfied if either $x - 1 = 0$ or $x - 3 = 0$.

But if $x - 1 = 0$, then $x = 1$, and if $x - 3 = 0$, then $x = 3$.

∴ equation (A) must be satisfied if either $x = 1$ or $x = 3$.

∴ 1, and 3, are the roots of the equation:

$$(x - 1)(x - 3) = 0,$$
i.e., $$x^2 - 4x + 3 = 0.$$

If it is required to solve the equation $x^2 - 4x + 3 = 0$, this can be done by reversing the above steps; consequently we find the factors of $x^2 - 4x + 3$ and so we get:

$$(x - 1)(x - 3) = 0.$$

The method is an easy one, if it is possible to obtain factors. In equations which arise out of practical work this can seldom be done.

It is a valuable method in other ways, and can be employed in equations of higher degree. If, for example, we know that

$$(x - 1)(x - 3)(x - 4) = 0.$$

Then by the above reasoning this equation is satisfied by:

$$x - 1 = 0, \text{ whence } x = 1$$
$$x - 3 = 0, \quad \text{,,} \quad x = 3$$
$$x - 4 = 0, \quad \text{,,} \quad x = 4.$$

The product of the three factors is an expression of the third degree, since the term of highest degree will be x^3. The equation is therefore of the third degree, or a **cubic** equation.

125. Worked examples.

Example 1. *Solve the equation $x^2 - 2x = 15$.*

Transposing

$$x^2 - 2x - 15 = 0.$$

Factorising

$$(x - 5)(x + 3) = 0.$$

$$\therefore \qquad\qquad x - 5 = 0 \text{ and } x = 5$$
or $\qquad\qquad x + 3 = 0 \text{ and } x = -3.$

$\therefore$ the solution is $x = 5$ or $x = -3$.

' **Note.**—The student should remember that the reasons for this method require that the right-hand side should be zero. Thus it cannot be used in such a case as

$$(x - 6)(x - 2) = 4.$$

Example 2. *Solve the equation $9x(x + 1) = 4$.*

Simplifying and transposing

$$9x^2 + 9x - 4 = 0.$$

Factorising

$$(3x + 4)(3x - 1) = 0.$$

$$\therefore \qquad\qquad 3x + 4 = 0 \text{ and } x = -\tfrac{4}{3}$$
or $\qquad\qquad 3x - 1 = 0 \text{ and } x = \tfrac{1}{3}.$

$\therefore$ the solution is $x = -\tfrac{4}{3}$ or $x = \tfrac{1}{3}$.

Exercise 36.

Solve the following equations by the method of factors:

1. $x(x - 3) = 0.$
2. $x(x + 5) = 0.$
3. $(x - 2)^2 = 0.$
4. $(x - 1)(x - 2) = 0.$
5. $(x + 4)(x - 1) = 0.$
6. $2(x - 3)(x + 7) = 0.$
7. $4(3x - 7)(2x + 11) = 0.$
8. $x^2 - 9x + 20 = 0.$
9. $x^2 + x = 6.$
10. $x^2 + 2x = 35.$
11. $x(x + 13) + 30 = 0.$
12. $x(x - 4) = x + 66.$
13. $(x - 8)(x + 4) = 13.$

14. $2x^2 - 11x + 12 = 0.$ 15. $2x^2 - 3x - 5 = 0.$
16. $3x^2 - 4x + 1 = 0.$ 17. $8x^2 - 14x - 15 = 0.$
18. $24x^2 + 10x - 4 = 0.$
19. $(x - 2)(x - 4)(x - 5) = 0.$
20. $(2x - 1)(x + 3)(x + 2) = 0.$
21. $(x - 1)(x^2 - 2x - 8) = 0.$
22. $(2x - 5)(x^2 - 5x - 50) = 0.$
23. $(x - a)(x - b) = 0.$ 24. $(2x - c)(x + d) = 0.$

126. General formula for the solution of a quadratic equation.

We have seen that by simplification and transposing every quadratic can be written in a form such as:

$$2x^2 + 7x - 4 = 0,$$

in which there are three constants—viz.:

    (1) the coefficient of x^2,
    (2)   ,,   ,, x,
    (3) the term independent of x.

Consequently if we want to write down a general form for any quadratic, letters such as a, b, c can be chosen to represent these constants, so that we could write the general form as:

$$ax^2 + bx + c = 0.$$

The equation above is a special case of this in which $a = 2$, $b = 7$, $c = -4$.

If we solved the general quadratic, $ax^2 + bx + c = 0$, the roots would be in terms of a, b, c. We should then have a formula such that, by substituting the values of a, b and c in any special case, we should be able to **write down the roots,** and no actual solving would be necessary.

127. Solution of the quadratic equation $ax^2 + bx + c = 0$.

We must use the method of " completing the square " as in § 121:

$$ax^2 + bx + c = 0.$$

Using preliminary step (1) $ax^2 + bx = -c.$

    ,,   ,,   (2) $x^2 + \dfrac{b}{a}x = -\dfrac{c}{a}.$

Adding " the square of half the coefficient of x "

$$x^2 + \frac{b}{a}x + \left(\frac{b}{2a}\right)^2 = \frac{b^2}{4a^2} - \frac{c}{a}.$$

$$\therefore \qquad \left(x + \frac{b}{2a}\right)^2 = \frac{b^2 - 4ac}{4a^2} \text{ (adding the fractions).}$$

Taking square roots

$$x + \frac{b}{2a} = \pm \frac{\sqrt{b^2 - 4ac}}{2a}.$$

$$\therefore \qquad x = -\frac{b}{2a} \pm \frac{\sqrt{b^2 - 4ac}}{2a}$$

or

$$x = \frac{-b \pm \sqrt{b^2 - 4ac}}{2a}$$

or in full

$$x = \frac{-b + \sqrt{b^2 - 4ac}}{2a}$$

or

$$x = \frac{-b - \sqrt{b^2 - 4ac}}{2a}.$$

This formula should be carefully learnt : by using it, there is no real necessity for working out any quadratic equation. Before applying it, however, the equation to be solved must be written down with all the terms on the left side.

From this formula, since there are always two square roots of a number, $\sqrt{b^2 - 4ac}$ must always have two values. Therefore x must have two values and **every quadratic equation must have two roots.** This fact was obvious in the solution by means of a graph.

128. Worked examples.

Example I. *Solve the equation* $5x^2 + 9x = 2$.

Writing the equation in general form we get:

$$5x^2 + 9x - 2 = 0.$$

Using the formula

$$x = \frac{-b \pm \sqrt{b^2 - 4ac}}{2a}$$

we have for this example

$$a = 5, \, b = 9, \, c = -2.$$

Substituting

$$x = \frac{-9 \pm \sqrt{9^2 - (4 \times 5 \times -2)}}{2 \times 5}$$

$$= \frac{-9 \pm \sqrt{81 + 40}}{10}$$

$$= \frac{-9 \pm \sqrt{121}}{10}$$

$$= \frac{-9 \pm 11}{10}$$

$$\therefore \qquad x = \frac{-9 + 11}{10} = \frac{2}{10}$$

$$\text{or} \qquad x = \frac{-9 - 11}{10} = \frac{-20}{10}.$$

$$\therefore \qquad x = 0 \cdot 2 \text{ or } -2.$$

Example 2. *Solve the equation* $\dfrac{1}{x-1} + \frac{2}{3} = \dfrac{2}{x-3}$.

Clearing fractions by multiplying throughout by $3(x-1)(x-3)$

$$3(x-3) + 2(x-1)(x-3) = 6(x-1)$$
$$3x - 9 + 2(x^2 - 4x + 3) = 6x - 6$$
$$3x - 9 + 2x^2 - 8x + 6 = 6x - 6.$$

Collecting like terms

$$2x^2 - 11x + 3 = 0.$$

Using the formula

$$x = \frac{-b \pm \sqrt{b^2 - 4ac}}{2a}$$

and putting $a = 2$, $b = -11$, $c = 3$,

we get
$$x = \frac{11 \pm \sqrt{121 - 24}}{4}$$

$$= \frac{11 \pm \sqrt{97}}{4} = \frac{11 + 9 \cdot 849 \text{ (app.)}}{4}.$$

$$\therefore \qquad x = \frac{11 + 9 \cdot 849}{4} = \frac{20 \cdot 849}{4} = 5 \cdot 212$$

$$\text{or} \qquad x = \frac{11 - 9 \cdot 849}{4} = \frac{1 \cdot 151}{4} = 0 \cdot 288.$$

$$\therefore \text{ the solution is } x = 5 \cdot 212 \text{ or } x = 0 \cdot 288.$$

Exercise 37.

Solve the following equations by using the formula of § 160:

1. $x^2 + 3x - 1 = 0.$

2. $x^2 - 5x + 2 = 0.$

3. $x^2 - 0\cdot4x = 1\cdot6.$

4. $3x^2 - 5x + 1 = 0.$

5. $2x^2 - 5x = 2.$

6. $5x^2 = 7x + 3.$

7. $2x^2 + 12x - 7 = 0.$

8. $3x^2 = 8x + 8.$

9. $4x^2 + 10x = 5.$

10. $4x(x + 2) = 9.$

11. $4x^2 = (x + 4)(2 - x).$

12. $0\cdot9(x + 1) = 0\cdot8 - x^2.$

13. $3x^2 = 7x + 2.$

14. $\dfrac{2}{3x + 1} + \dfrac{1}{2x + 1} = \tfrac{2}{7}.$

15. $0\cdot5 + \dfrac{2}{x + 3} - \dfrac{11}{x + 7} = 0.$

16. $\dfrac{3}{x + 2} - \dfrac{1}{x - 3} = \dfrac{4}{x}.$

17. $(x + 3)(x - 5) = 2x - 2.$

---

129. Problems leading to quadratics.

The following examples illustrate the method of solving such problems.

Example 1. *The distance (h) which a body reaches in time t when it is projected vertically upwards with velocity u is given by the formula*

$$h = ut - \tfrac{1}{2}gt^2.$$

If u = 160 and g = 10, find the time a body takes to rise 240.

Substituting the given values in the formula

$$240 = 160t - 5t^2$$
$$5t^2 - 160t + 240 = 0$$
$$t^2 - 32t + 48 = 0$$

$$\therefore \qquad t = \frac{32 \pm \sqrt{32^2 - 4 \times 48}}{2}$$

$$= 16 \pm 4\sqrt{13}$$
$$= 16 \pm 3\cdot6 \text{ approx.}$$
$$t = 19\cdot6 \text{ or } 12\cdot4.$$

The two roots require consideration. Since every quadratic has two roots, when these furnish the answer to a problem the applicability of the roots to the problem must be examined. Sometimes it will be clear that both are not applicable, especially if one is negative. Considering the above problem, when a body is projected vertically upwards its velocity decreases until it reaches its highest point, when it is zero. It then falls vertically and retraces its path. Therefore **it will be at a given height twice, once when ascending and again when descending.** The value $t = 1·84$ gives the time to reach 204 ft. when going up, and it is at the same height when descending $8·16$ secs. after starting.

Example 2. *A motorist travels a distance of 84 km. He finds that, if on the return journey he increases his average speed by 4 km/h, he will take half an hour less. What was his average speed for the first part of the journey and how long did he take for the double journey?*

Let x miles per hour be the average speed for the first journey.

Then *time* for the first journey is $\dfrac{84}{x}$ hours.

Speed for the return journey is $(x + 4)$ miles per hour.

$\therefore$ *time* for the second journey is $\dfrac{84}{x + 4}$.

But this is $\frac{1}{2}$ hour less than the first time.

$\therefore \qquad\qquad \dfrac{84}{x} - \dfrac{84}{x + 4} = \tfrac{1}{2}.$

Clearing fractions by multiplying by the L.C.D.—viz., $2x(x + 4)$.

$$168(x + 4) - 168x = x(x + 4)$$
$$168x + 672 - 168x = x^2 + 4x.$$

Collecting and transposing

$$x^2 + 4x - 672 = 0.$$

Using the formula

$$x = \frac{-b \pm \sqrt{b^2 - 4ac}}{2a}$$

or substituting

$$x = \frac{-4 \pm \sqrt{16 - (4 \times -672)}}{2}$$

$$= \frac{-4 \pm \sqrt{2704}}{2}$$

$$= \frac{-4 \pm 52}{2}.$$

$$\therefore \quad x = \frac{-4 + 52}{2} = 24$$

or $$x = \frac{-4 - 52}{2} = -28.$$

The negative root, although it satisfies the equation above, has no meaning for this problem.

$\therefore$ average speed for first journey is 24 km/h.

Time for first journey $= \frac{84}{24} = 3\frac{1}{2}$ h.

,, second ,, $= \frac{84}{28} = 3$ h.

$\therefore$ total time $= 6\frac{1}{2}$ hours.

Exercise 38.

1. The sum of a number and its reciprocal is 2·9. Find the number.

2. The area of a rectangle is 135 mm² and its perimeter is 48 mm. What are the lengths of its sides?

3. Solve for $\frac{1}{R}$ the equation $\frac{1}{R^2} - \frac{5}{R} = 10$.

4. The relation between the joint resistance R and two resistances r_1 and r_2 is given by the formula $\frac{1}{R} = \frac{1}{r_1} + \frac{1}{r_2}$. If R is 12 ohms and r_2 is 6 ohms greater than r_1, find r_1 and r_2.

5. A formula for finding the strength of a concrete beam is $bn^2 + 2am(n - c) = 0$. Solve this for n when $b = 4$, $a = 2$, $c = 8$, $m = 12·5$.

6. The formula giving the sag (D) in a cable of length L and span S is expressed by $L = \frac{8D^2}{3S} + S$. Find S when $L = 80$, $D = 2·5$.

7. The product of a number n and $2n - 5$ is equal to 250. What is the value of n?

8. There is an algebraical formula $s = \frac{n}{2}\{2a + (n-1)d\}$. If $s = 140$, $a = 7$ and $d = 3$, find n.

9. By a well-known geometrical theorem the square on the diagonal of a rectangle is equal to the sum of the squares on the two sides. The diagonal of a particular rectangle is 25 mm long, and one side of the rectangle is 5 mm longer than the other. Find the sides of the rectangle.

10. The cost of a square carpet is £24·80 and the cost of another square carpet whose side is 3 m longer than that of the first is £38·75. If the cost per m² is the same for both carpets, find the area of each.

11. There is a number such that when it is increased by 3 and the sum is squared, the result is equal to 12 times the number increased by 16. What is the number?

12. Two adjacent sides of a rectangle are represented by $(x + 4)$ and $(x + 6)$. The area of the rectangle is equal to twice the area of the square whose side is x. What is the value of x?

13. Find the area of a rectangular plot of ground whose perimeter is 42 m and whose diagonal is 15 m.

14. The cost of boring a well is given by the formula, $C = 2x + \frac{1}{50}x^2$, where C is the cost in pounds and x is the depth in metres. If a well cost £280 to bore, how deep was it?

15. One number exceeds another by 4. The sum of their squares is 208. What are the numbers?

16. The formula for the sum of the first n whole numbers is $\frac{1}{2}n(n + 1)$. If the sum is 78, how many numbers are there?

---

130. Simultaneous equations of the second degree.

These are equations which involve two unknowns and which include terms of the second degree.

They are also called **simultaneous quadratics**.

The **degree of a term** is shown either by its index, or, if it contains two or more letters, by the sum of their indices.

x^2, xy, y^2 are terms of the second degree involving two unknowns, and any of these, together with terms of the first degree, may occur in the equations to be solved. Numerical coefficients do not affect the " degree " of a term.

It is seldom possible to solve simultaneous quadratics unless they conform to certain types. In this book we shall consider two of these types.

131. First type. When one of the equations is of the first degree.

Example.

$$x + y = 1 \quad . \quad . \quad . \quad . \quad (1)$$
$$3x^2 - xy + y^2 = 37 . \quad . \quad . \quad . \quad (2)$$

It is always possible to solve simultaneous quadratics if we can find one letter in terms of the other, the relation being a linear one.

In the equations above, from (1) we get:

$$y = 1 - x.$$

This can now be substituted in equation (2).
Thus we get:

$$3x^2 - x(1 - x) + (1 - x)^2 = 37.$$

In this way a quadratic, with one unknown, is reached, and this can always be solved by the methods previously given.

Simplifying

$$3x^2 - x + x^2 + 1 - 2x + x^2 = 37.$$
$$\therefore \qquad 5x^2 - 3x - 36 = 0.$$

This could be solved by factorisation, or using the formula :

$$x = \frac{-b \pm \sqrt{b^2 - 4ac}}{2a}.$$

On substituting $x = \dfrac{+3 \pm \sqrt{9 + 720}}{10}$

$$= \frac{3 \pm 27}{10}.$$

$$\therefore \qquad x = \frac{30}{10} \text{ or} - \frac{24}{10}$$

or $\qquad x = 3 \text{ or} - 2\cdot4.$

To find y substitute in

$$y = 1 - x.$$

(1) *When* $x = 3$,

$$y = 1 - 3 = -2.$$

(2) *When* $x = -2.4$,

$$y = 1 - (-2.4) = 3.4.$$

∴ the solutions are:

(1) $x = 3, y = -2.$
(2) $x = -2.4, y = 3.4.$

The solutions should be arranged in corresponding pairs.

Exercise 39.

Solve the following equations:

1. $x - y = 2.$
   $x^2 + xy = 60.$
2. $x + y = 7.$
   $3x^2 + xv - v^2 = 81.$
3. $2x + y = 5.$
   $5x^2 - 3xy = 14.$
4. $3x + y = 8.$
   $x^2 - 5xy + 8y = 36.$
5. $x + y + 1 = 0.$
   $3x^2 - 5y^2 - 7 = 0.$
6. $2x + 3y = 14.$
   $4x^2 + 2xy + 3y^2 = 60.$
7. $2x^2 - 5x + 4xy = 60.$
   $3x - y = 9.$
8. $2x + 3y = 5.$
   $\dfrac{5}{y + 3} - \dfrac{9}{x} = 1.$
9. $3x + y = 25.$
   $xy = 28.$
10. $2x - y = 8.$
    $\dfrac{x^2}{9} + \dfrac{y^2}{16} = 5.$
11. $\dfrac{1}{x} + \dfrac{1}{y} = \frac{3}{4}.$
    $3x - y = 2.$

---

132. Second type. Symmetric equations.

Example I. *An example of this type is*

$$x + y = 19 \quad . \quad . \quad . \quad . \quad . \quad (1)$$
$$xy = 84 \quad . \quad . \quad . \quad . \quad (2)$$

These two equations are called " symmetric " because, if the letters x and y are interchanged throughout, the equations are unaltered. Other equations which are not strictly symmetric can be solved by the same method as these, and so are included.

In these particular equations number (1) is of the first degree, and so the previous method can be used; but this is an easy example to illustrate the special solution which can be employed with this type.

The aim of the method is to find the value of $x - y$. Then, since the value of $x + y$ is known, the rest of the solution is easy.

We are given

$$x + y = 19 \quad \cdots \quad (1)$$
$$xy = 84 \quad \cdots \quad (2)$$

Squaring both sides of (1)

$$x^2 + 2xy + y^2 = 361 \quad \cdots \quad (3)$$

From (2)

$$4xy = 336 \quad \cdots \quad (4)$$

Subtracting (4) from (3) we get

$$x^2 - 2xy + y^2 = 25.$$

∴ $\qquad\qquad x - y = \pm 5.$

From (1) $\qquad\qquad x + y = 19.$

Adding $\qquad\qquad 2x = 19 \pm 5 = 24 \text{ or } 14.$

∴ $\qquad\qquad x = 12 \text{ or } 7.$

Subtracting $\qquad\qquad 2y = 19 - (\pm 5)$
$\qquad\qquad\qquad\qquad = 14 \text{ or } 24.$

∴ $\qquad\qquad y = 7 \text{ or } 12.$

Or using $\qquad\qquad xy = 84$

(1) when $\qquad\qquad x = 12, y = 7,$

(2) ,, $\qquad\qquad x = 7, y = 12.$

The symmetry of the two equations appears in the solution.

Example 2. *Solve the equations*:

$$x^2 + y^2 = 89 \quad \cdots \quad (1)$$
$$xy = 40 \quad \cdots \quad (2)$$

Both equations are of the second degree. The aim in this case is to obtain both $x + y$ and $x - y$.

From (2)

$$2xy = 80 \quad \cdots \quad (3)$$

Adding (1) and (3)

$$x^2 + 2xy + y^2 = 169.$$
$$\therefore \qquad x + y = \pm 13.$$

Subtracting (3) from (1)

$$x^2 - 2xy + y^2 = 9.$$
$$\therefore \qquad x - y = \pm 3.$$

Associating the equations:

$$x + y = \pm 13.$$
$$x - y = \pm 3.$$

Adding $\quad 2x = \pm 13 \pm 3 = \pm 16$ or $\pm 10.$

$\therefore \qquad x = \pm 8$ or $\pm 5.$

To get corresponding values of x we may now substitute in

$$xy = 40.$$

$\therefore$ when $\quad x = + 8, - 8, + 5, - 5$
$$y = + 5, - 5, + 8, - 8.$$

$\therefore$ associating these the solution is:

$$x = + 8, y = + 5.$$
$$x = - 8, y = - 5.$$
$$x = + 5, y = + 8.$$
$$x = - 5, y = - 8.$$

Note.—In the above examples the sum and difference of two numbers, $x + y$, $x - y$, are used. The same method would be used if we were given such an example as $x + 3y$. We should then aim at getting $(x + 3y)^2$, $(x - 3y)^2$, and ultimately $x - 3y$. There are many other variations in this type of solution. In some, such as

$$x^2 + y^2 = a \quad . \quad . \quad . \quad . \quad (1)$$
$$x + y = b \quad . \quad . \quad . \quad . \quad (2)$$

xy is not given, but it can be obtained by squaring equation (2) and subtracting equation (1).

Exercise 40.

Solve the following equations:

1. $x + y = 8.$
   $xy = 15.$

2. $x + y = 9.$
   $3xy - 42 = 0.$

3. $x - y = 5.$
   $xy = 24.$

4. $x^2 + y^2 = 17.$
   $x + y = 5.$

5. $x - 2y = 2.$
   $xy = 12.$

6. $x^2 + y^2 = 17.$
   $xy = 4.$

7. $x^2 - xy + y^2 = 67.$
   $xy = 18.$

8. $x^2 - xy + y^2 = 43.$
   $x + y = 8.$

9. $x^2 + y^2 = 45.$
   $x^2 - xy + y^2 = 27.$

10. $2x + 3y = 9.$
    $xy = 3.$

CHAPTER XV

INDICES

133. The meaning of an index.

In § 18 it was shown that the product of a number of equal factors such as $a \times a \times a \times a$ can be written in the form a^4, in which 4 is an index which indicates the number of factors.

Generalising this, if there be n such equal factors, where n is a positive integer, then a^n will be defined as follows:

$$a^n = a \times a \times a \times \ldots \text{ to } n \text{ factors}$$
and a^n is called the nth power of a.

134. Laws of indices.

It was also shown in §§ 19, 20, 21 that when operations such as multiplication and division of powers of a number are performed, the laws which govern these operations can be deduced from the definitions of a power and an index given in § 133.

General proofs of these laws will now be given.

I. First law of indices: the law of multiplication.

The special cases given in § 19 lead to the general law as follows:

Let m and n be any positive integers.

By definition

$$a^m = a \times a \times a \times \ldots \text{ to } m \text{ factors}$$
and $a^n = a \times a \times a \times \ldots \text{ to } n \text{ factor.}$

Then $a^m \times a^n = (a \times a \times a \times \ldots \text{ to } m \text{ factors}) \times (a \times a \times a \times \ldots \text{ to } n \text{ factors}).$

But when two groups of factors are multiplied the factors in the groups are associated as one group of factors to give the product (see § 19).

$\therefore a^m \times a^n = a \times a \times a \times a \times \ldots$ to $(m + n)$ factors
$= a^{m + n}$ (by definition).

$\therefore$ the first law of indices is

$$a^m \times a^n = a^{m+n}$$

The law is clearly true when the product involves more than two powers. Thus $a^m \times a^n \times a^p = a^{m + n + p}$.

II. Second law of indices: the law of division.

To find the value of $a^m \div a^n$.

With the same definition as before, and proceeding as in the special cases of § 21.

$$a^m \div a^n = \frac{a \times a \times a \times \ldots \text{ to } m \text{ factors}}{a \times a \times a \times \ldots \text{ to } n \text{ factors}}.$$

After cancelling the n factors of the denominator with n corresponding factors in the numerator, there are left in the numerator $m - n$ factors.

$$\therefore a^m \div a^n = a^{m - n}$$

Note.—This proof assumes that m is greater than n. If m is less than n, there are $n - m$ factors left in the denominator.

$$\therefore \text{ if } n > m \qquad a^m \div a^n = \frac{1}{a^{n - m}}.$$

This case will be examined later.

III. Third law of indices: the law of powers.

By the definition of § 133 such an expression as $(a^4)^3$ means the third power of a^4, or

$$\begin{aligned}
(a^4)^3 &= a^4 \times a^4 \times a^4 \\
&= a^{4 + 4 + 4} \quad \text{(by first index law)} \\
&= a^{4 \times 3} \\
&= a^{12}.
\end{aligned}$$

In general, if m and n are any positive integers by definition

$$\begin{aligned}
(a^m)^n &= a^m \times a^m \times a^m \ldots \text{ to } n \text{ factors} \\
&= a^{m + m + m} \cdots \text{ to } n \text{ terms (first index law)} \\
&= a^{m \times n}.
\end{aligned}$$

∴ the law of powers is

$$(a^m)^n = a^{m \times n}.$$

For powers of a product such as $(ab)^n$, see § 20. In a similar way it may be shown that

$$(ab)^n = a^n \times b^n.$$

Exercise 41.

Revision Exercises in Indices.

Write down the values of the following:

1. $2a^4 \times 3a^5$.
2. $a^6 \times a^6$.
3. $\frac{1}{2}a^5 \times \frac{1}{4}a^3$.
4. $2 \times 2^2 \times 2^3 \times 2^4$.
5. $4 \times 4^2 \times 4^3$.
6. $3a^3b^4 \times 2a^5b^3$.
7. $x^4y^5z \times x^2y^7z^3$.
8. $x^{m+1} \times x^{m-1}$.
9. $a^{m+n} \times a^{m-n}$.
10. $a^pb^q \times a^{p+q} \times b^{p-q}$.
11. $a^{n+3} \times a^{n-3}$.
12. $a^m \times a^n \times a^4$.

---

13. $a^8 \div a^5$.
14. $5x^6 \div 10x^3$.
15. $36a^{10} \div 12a^5$.
16. $2^{10} \div 2^4$.
17. $3^5 \div 3^2$.
18. $(-x^7) \div (-x^2)$.
19. $15a^4b^2 \div -3ab$.
20. $a^{2p} \div a^p$.
21. $12x^{2p}y^{2q} \div 3x^py^q$.
22. $\dfrac{a^{10} \times a^4}{a^7}$.
23. $\dfrac{x^8}{x^4} \times \dfrac{x^3}{x^6}$.
24. $a^{2n} \div a^{n-1}$.
25. $a^{m+n} \div a^{m-n}$.
26. $a^{n+4} \div a^{n-2}$.
27. $\dfrac{4a^3}{3b^2} \div \dfrac{6a^2}{4b^3}$.
28. $\dfrac{a}{b} \div \dfrac{b}{a}$.

---

29. $(2^3)^2$.
30. $(3^2)^3$.
31. $(x^5)^2$.
32. $(x^2)^5$.
33. $(a^4)^4$.
34. $(2a^2b^3)^2$.
35. $\left(\dfrac{x^4y^5}{2y^2}\right)^3$.
36. $(a^p)^3$.
37. $(x^4)^n$.
38. $(3a^{2p})^3$.
39. $\sqrt{a^4}$.
40. $\sqrt{x^8}$.
41. $\sqrt{x^{16}}$.
42. $\sqrt{9a^6}$.

43. $\sqrt{a^4b^2}$.

44. $\sqrt{\dfrac{x^6}{y^4}}$.

45. $\sqrt{\dfrac{9a^8}{4b^6}}$.

46. $\sqrt{x^{2n}}$.

---

135. Extension of the meaning of an index.

It has so far been assumed that all indices are positive integers. The definition of a^n—viz.:

$$a^n = a \times a \times a \times \ldots \text{ to } n \text{ factors}$$

is unintelligible except upon the assumption that n is a positive integer.

But Algebra generalises, and we must therefore consider the possibility of attaching a meaning to an index in all cases.

136. Graph of 2^x.

As a first step let us choose a suitable number, say 2, and plot some of the powers of it—*i.e.*, draw the graph of 2^x, in which x represents any index. Calculating the values of these powers for some of the smaller integral values of x, we get a table of values as follows:

x . .	1	2	3	4
$y = 2^x$.	2	4	8	16

When these points are plotted they appear to be on a smooth curve, as drawn in Fig. 40.

If we are justified in assuming that this curve is continuous and such that all the points on it satisfy the equation $y = 2^x$, then it follows that if any point be taken on the curve between the plotted points, its co-ordinates must also satisfy the law.

Therefore if any point, A, be taken on the curve at which $x = 1\cdot5$ and $y = 2\cdot8$, then it follows that $2^{1\cdot5} = 2\cdot8$.

Again at B, where $x = 3\cdot2$ and $y = 9$, $2^{3\cdot2} = 9$.

Hence, if the assumption is correct, that the curve is a continuous one, and the co-ordinates of any point on it satisfy $y = 2^x$, then it may be concluded that **any number** within the limits plotted can be expressed as a power of 2, and conversely any number can be used as the index of some power of 2.

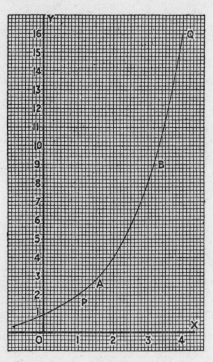

Fig. 40.

A similar curve could be drawn exhibiting the powers of any other number—say, 10—*i.e.*, we could draw the curve of
$$y = 10^x.$$

Thus we could express any number as a power of 10.

Reverting to the curve PQ (Fig. 40), since it was plotted by taking values of x from unity, the curve began at P. If it be produced towards the y axis in the way it seems to curve, it will apparently cut the y axis at the point where $y = 1$.

This suggests that:

(1) The value of 2^x when $x = 0$ is 1—*i.e.*, $2^0 = 1$.
(2) The portion of the curve thus drawn should show values of 2^x between $x = 0$ and $x = 1$.

For example, when $x = \frac{1}{2}$, 2^x is approximately 1·4—*i.e.*, $2^{\frac{1}{2}} = 1·4$ approximately.

Again, the curve evidently does not end at the y axis, but can be further produced. This portion will correspond to negative values of x. Hence we infer that values of 2^x can be found when the index is negative.

137. Algebraical consideration of the extension of the meaning of indices.

From the graph of powers it may be inferred that powers of a number exist whether the index be integral or fractional, positive or negative. We must now consider how such indices can be interpreted algebraically.

In seeking to find meanings for the new forms of indices we must be guided by one fundamental principle—viz.:

Every index must obey the laws of indices as discovered for positive integral numbers.

Otherwise it cannot be considered as an index.

138. Fractional indices.

We will begin with the simple case of $a^{\frac{1}{2}}$.

To find a meaning for $a^{\frac{1}{2}}$.

Since $a^{\frac{1}{2}}$ must conform to the laws of indices
∴ by the first law

$$a^{\frac{1}{2}} \times a^{\frac{1}{2}} = a^{\frac{1}{2} + \frac{1}{2}}$$
$$= a^1 \text{ or } a.$$

∴ $a^{\frac{1}{2}}$ must be such a number that on being multiplied by itself the product is a.

But such a number is defined arithmetically as the **square root** of a.

∴ $a^{\frac{1}{2}}$ must be defined as $\sqrt{a}$.

As an example $2^{\frac{1}{2}} = \sqrt{2} = 1·414$ approx.
This agrees with the value of 1·4 found from the graph

(Fig. 40). The reasoning above clearly holds in all such cases, and so we may infer that in general, if n be any positive integer

$$a^{\frac{1}{n}} = \sqrt[n]{a}.$$

To find a meaning for $a^{\frac{2}{3}}$.

Applying the first law of indices:

$$a^{\frac{2}{3}} \times a^{\frac{2}{3}} \times a^{\frac{2}{3}} = a^{\frac{2}{3} + \frac{2}{3} + \frac{2}{3}}$$
$$= a^2.$$
$$\therefore \qquad a^{\frac{2}{3}} = \sqrt[3]{a^2}.$$

Applying the reason generally, we can deduce that if m and n are any positive integers

$$a^{\frac{m}{n}} = \sqrt[n]{a^m}.$$

Thus
$$a^{\frac{3}{4}} = \sqrt[4]{a^3}$$
$$a^{\frac{2}{5}} = \sqrt[5]{a^2}.$$

Indices which are in decimal form can be changed to vulgar fractions. Thus:

$$a^{0 \cdot 25} = a^{\frac{1}{4}}$$
$$= \sqrt[4]{a}.$$

139. To find a meaning for a^0.

Using the second law of indices, and n being any number,

$$a^n \div a^n = a^{n-n}$$
$$= a^0.$$
But
$$a^n \div a^n = 1.$$
$$\therefore \qquad a^0 = 1.$$

This confirms the conclusion reached in § 136 from the graph. It should be noted that a represents any finite number. Therefore $a^0 = 1$, whatever the value of a.

Graphically, if curves of a^x are drawn for various values of a, as that of 2^x was drawn in Fig. 40, all these curves will pass through the point on the y axis which is unit distance from 0.

140. Negative indices.

In considering the curve of 2^x in Fig. 40, the conclusion

was reached that the curve could be continued for
negative values of x. We must now find what meaning
can be given algebraically to a negative index.

Consider a^{-1}.

This must obey the laws of indices.

∴ by the first law

$$a^{-1} \times a^{+1} = a^{-1+1}$$
$$= a^0$$
$$= 1.$$

Dividing by a^{+1} we get

$$a^{-1} = \frac{1}{a},$$

i.e., by a^{-1} we must mean " the reciprocal of a ".

With the same reasoning

$$a^{-2} = \frac{1}{a^2}$$

$$a^{-3} = \frac{1}{a^3},$$

and in general

$$a^{-n} \times a^{+n} = a^{-n+n}$$
$$= a^0$$
$$= 1.$$

∴ $$a^{-n} = \frac{1}{a^n}.$$

∴ a^{-n} is defined as the reciprocal of a^n.

The following examples should be noted:

$$2a^{-3} = \frac{2}{a^3} \quad \text{(negative index applies to } a \text{ only)}$$

$$a^{-\frac{1}{2}} = \frac{1}{\sqrt{a}}$$

$$\frac{1}{a^{-1}} = a$$

and generally $\quad \dfrac{1}{a^{-n}} = a^n.$

It should be noted as a working rule that when a power
of a number is transferred from the numerator of a fraction
to the denominator, or vice-versa, **the sign of the index is
changed.**

If n is a positive number a^n increases as n increases.

$\therefore$ a^{-n} or $\dfrac{1}{a^n}$ decreases as n increases.

Consequently the curve of 2^x will approach closer to the x axis for negative values of x.

This agrees with the course which the curve in Fig. 40 appeared to be taking.

141. Standard forms of numbers.

Indices may be usefully employed in what are called **standard forms,** by means of which we can express clearly and concisely certain numbers which are very large or very small.

The method of formation of them will be seen from the following examples:

$$26 = 2 \cdot 6 \times 10^1$$
$$260 = 2 \cdot 6 \times 10^2$$
$$2600 = 2 \cdot 6 \times 10^3$$
$$26\,000 = 2 \cdot 6 \times 10^4$$
$$260\,000 = 2 \cdot 6 \times 10^5 \quad \text{and so on.}$$

When a number is written in standard form, **one** digit only is retained in the whole number part, and this is multiplied by a power of 10 to make it equal to the number. It should be noted that **the index of the power of 10 is equal to the number of figures which follow the figure retained in the integral part of the number.**

Thus in 547 000 000, when written in standard form, **5** only is retained in the whole number part and 8 figures follow it.

$\therefore$ $\qquad\qquad$ $547\,000\,000 = 5 \cdot 47 \times 10^8.$

Numbers which are less than unity may be similarly changed into standard form by using negative indices.

Thus $\quad$ $0 \cdot 26$ $\quad = 2 \cdot 6 \div 10$ or $2 \cdot 6 \times 10^{-1}$
$\qquad\quad$ $0 \cdot 026$ $\quad = 2 \cdot 6 \div 10^2$ or $2 \cdot 6 \times 10^{-2}$
$\qquad\quad$ $0 \cdot 0026$ $\quad = 2 \cdot 6 \div 10^3$ or $2 \cdot 6 \times 10^{-3}$
$\qquad\quad$ $0 \cdot 000\,26 = 2 \cdot 6 \div 10^4$ or $2 \cdot 6 \times 10^{-4}$, etc.

It should be noted that **the numerical part of the index is one more than the number of zeros following the decimal point in the original number.**

142. Operations with standard forms.

In these operations the rules of indices must be observed.

Examples.

(1) $(1 \cdot 2 \times 10^4) \times (2 \cdot 3 \times 10^3) = (1 \cdot 2 \times 2 \cdot 3) \times (10^4 \times 10^3)$
$$= 2 \cdot 76 \times 10^7.$$

(2) $(4 \cdot 8 \times 10^8) \div (1 \cdot 6 \times 10^{-3}) = (4 \cdot 8 \div 1 \cdot 6) \times (10^8 \div 10^{-3})$
$$= 3 \times 10^{11}.$$

Exercise 42.

Note.—When necessary in the following examples take $\sqrt{2} = 1 \cdot 414$, $\sqrt{3} = 1 \cdot 732$, $\sqrt{5} = 2 \cdot 236$, $\sqrt{10} = 3 \cdot 162$.

1. State what meanings can be given to the following:

(1) $3^{\frac{1}{2}}$. (2) $8^{\frac{1}{3}}$. (3) $a^{\frac{1}{2}}$.
(4) $2^{0 \cdot 5}$. (5) $3^{0 \cdot 25}$. (6) $a^{0 \cdot 75}$.

2. State what meanings can be given to the following:

(1) $10^{\frac{2}{3}}$. (2) $4^{\frac{3}{2}}$. (3) $a^{\frac{1}{4}}$.
(4) $10^{1 \cdot 5}$. (5) $2^{2 \cdot 5}$. (6) $10^{1 \cdot 25}$.

3. Write down as simply as possible the meaning of:

(1) $4^{\frac{3}{2}}$. (2) $8^{\frac{2}{3}}$. (3) $a^{1 \cdot 75}$.
(4) $a^{\frac{1}{2}}$. (5) $a^{1 \cdot 2}$. (6) $a^{0 \cdot 2}$.

4. Find the numerical values of:

(1) $4^{1 \cdot 5}$. (2) $16^{1 \cdot 25}$. (3) $9^{1 \cdot 5}$.
(4) $100^{\frac{3}{2}}$. (5) $10^{\frac{2}{3}}$. (6) $(\frac{1}{4})^{0 \cdot 5}$.

5. Find the values of 3^{-2}, 3^{-1}, 3^0, 3^1, $3^{1 \cdot 5}$, 3^2, $3^{2 \cdot 5}$. Plot these and draw the curve which contains the points.

6. Find the values of:

(1) $2^2 \times 2^{\frac{1}{2}}$. (2) $3 \times 3^{\frac{1}{2}} \times 3^{\frac{1}{2}}$.
(3) $10^{\frac{1}{2}} \div 10^{\frac{1}{2}}$. (4) $a^{\frac{1}{2}} \times a^{\frac{1}{2}}$.
(5) $2^{\frac{2}{3}} \div 2^{\frac{1}{3}}$. (6) $a^{1 \cdot 5} \div a^{0 \cdot 3}$.

7. Write down the meanings with positive indices of:

$$4^{-1},\ 3a^{-2},\ 2^{-\frac{1}{2}},\ \frac{1}{3^{-1}},\ \frac{3}{a^{-2}},\ 10^{-3}.$$

8. Write with positive indices:

(1) x^{-3}.   (2) $x^{-\frac{1}{2}}$.   (3) $\dfrac{1}{x^{-3}}$.

(4) $(x^{-3})^2$.   (5) $\left(\dfrac{1}{x}\right)^{-1}$.   (6) $\dfrac{1}{2x^{-\frac{1}{2}}}$.

9. Find the values of:

(1) $8^{\frac{2}{3}}$.   (2) $25^{\frac{3}{2}}$.   (3) $(10^2)^{\frac{3}{2}}$.

(4) $(5^{-3})^2$.   (5) $\dfrac{2}{2^{-3}}$.   (6) $81^{\frac{1}{4}}$.

10. Find the values of:

(1) $(\frac{1}{2})^{-2}$.   (2) $(\frac{2}{3})^{-3}$.   (3) $16^{-0.5}$.
(4) $(36)^{-0.5}$.   (5) $(4)^{1.5}$.   (6) $(\frac{1}{4})^{2.5}$.

11. Find the value of $a^4 \times a^{-2} \times a^{\frac{1}{2}}$ when $a = 2$.

12. Write down the simplest forms of:

(1) $a^{\frac{1}{2}} \times a^{\frac{1}{2}}$.   (2) $10^3 \times 10^{-\frac{1}{2}}$.
(3) $a^3 \div (-a)^6$.   (4) $-a^3 \div (-a)^5$.
(5) $x^{2n} \div x^{-n}$.   (6) $x^n \div x^{-n}$.

13. Find the values of:

(1) $a^3 \times a^0$.   (2) $a^0 \times 1$.   (3) $a^0 \times 0$.

14. Write down in " standard form " the answers to the following:

(1) $(2.2 \times 10^5) \times (1.6 \times 10^4)$.
(2) $(7.1 \times 10^3) \times (2.3 \times 10^3)$.
(3) $(4.62 \times 10^5) \div (2.1 \times 10^3)$.
(4) $(7.4 \times 10^6) \times (5 \times 10^{-4})$.
(5) $(1.2 \times 10^{-3}) \times (2.1 \times 10^{-4})$.

CHAPTER XVI

LOGARITHMS

143. A system of Indices.

In the previous chapter it was seen that by means of the graph of 2^x it is possible, within the limits of the graph, to express any number as a power of 2. This was confirmed algebraically.

For every number marked on the y axis and indicated on the graph, there is a corresponding index which can be read on the x axis. These constitute a system of indices by which numbers can be expressed as powers of a common basic number 2.

Similarly, by drawing graphs such as 3^x, 5^x, 10^x, numbers can be expressed as powers of 3, 5, or 10, or any other basic number.

Thus in all such cases it is possible to formulate systems of indices which, for any number A, would enable us to determine what power that number is of any other number B, which is called the base of the system.

This possibility of expressing any number as a power of any other number, and thus of the formation of a system of indices, as stated above, leads to practical results of great importance. It enables us to carry out, easily and accurately, calculations which would otherwise be laborious or even impossible. The fundamental ideas underlying this can be illustrated by means of a graph of powers similar to that drawn in Fig. 40. For this purpose we will use 10 as the base of the system and draw the graph of $y = 10^x$.

As powers of 10 increase rapidly, it will be possible to employ only small values for x, if the curve is to be of any use for our purpose. To obtain those powers we must use the rules for indices which were formulated in the previous chapter.

From Arithmetic we know that $\sqrt{10} = 3 \cdot 16$ app., *i.e.*, $10^{\frac{1}{2}} = 3 \cdot 16$.

Then $10^{0.25} = 10^{\frac{1}{4}} = (10^{\frac{1}{2}})^{\frac{1}{2}} = (3.16)^{\frac{1}{2}} = 1.78$ app. (by Arithmetic).

$10^{0.75} = 10^{\frac{3}{4}} = 10^{\frac{1}{2} + \frac{1}{4}} = 10^{\frac{1}{2}} \times 10^{\frac{1}{4}} = 3.16 \times 1.78$
$= 5.62$ app.

$10^{0.125} = 10^{\frac{1}{8}} = (10^{\frac{1}{4}})^{\frac{1}{2}} = (1.78)^{\frac{1}{2}} = 1.33$ app.

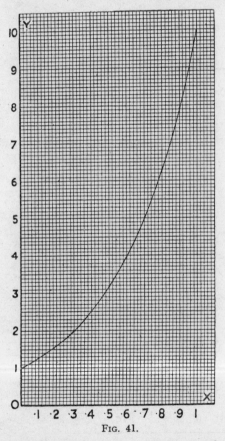

Fig. 41.

In this way a table of values for the curve can be compiled as follows:

x	0	0·125	0·25	0·5	0·75	0·875	1
10^x	1	1·33	1·78	3·16	5·62	7·5	10

The resulting curve is shown in Fig. 41. The following examples illustrate the use that can be made of it in calculations.

Example 1. *Find from the graph the value of* $1·8 \times 2·6$.

From the graph

$$1·8 = 10^{0·26}$$
$$2·6 = 10^{0·42}$$
$$\therefore \quad 1·8 \times 2·6 = 10^{0·26} \times 10^{0·42}$$
$$= 10^{0·26 + 0·42} \quad \text{(first law of indices)}$$
$$= 10^{0·68}$$
$$= 4·6 \quad \text{from the graph.}$$

Example 2. *Find* $\sqrt[3]{9}$.

From the graph

$$9 = 10^{0·96}$$
$$\sqrt[3]{9} = 9^{\frac{1}{3}}$$
$$= (10^{0·96})^{\frac{1}{3}}$$
$$= 10^{0·32} \quad \text{(third law of indices)}$$
$$= 2·1 \quad \text{from the graph.}$$

144. A system of logarithms.

Although interesting as illustrating the principles involved, the above has little practical value for purposes of calculation, since we depend upon the readings from a curve which is necessarily limited in size and is not sufficiently accurate.

For practical purposes it is obvious that tables of the indices used, calculated to a suitable degree of accuracy, are necessary. Such tables have been compiled and are available for the purpose. For their compilation a more advanced knowledge of mathematics is required than is included in this volume.

The tables are constructed with 10 as a suitable base. They give the indices which indicate for all numbers, within the scope of the table, the powers they are of 10.

Such a table is called a **system of logarithms,** and the number 10, with respect to which the logarithms are calculated, is called the **base of the system.**

A logarithm to base 10 may be defined as follows:

The logarithm of a number to base 10 is the index of the power to which 10 must be raised to produce the number.

145. Notation for logarithms.

The student may wonder why another, and an unfamiliar term is employed as a name for an index. One reason for this will be seen from the following:

Let n be any positive number.

,, x be its index to base 10.

Then $n = 10^x$.

This is in reality a formula. If it is required to " *change the subject of the formula* " (see § 51) and express x in terms of the other letters, there is a difficulty in doing this concisely. Using words we could write:

x = index of power of n to base 10.

This is cumbersome, so we employ the word " logarithm ",[1] abbreviated to " log " as follows:

$$x = \log_{10} n$$

the number indicating the base being inserted as shown.

If the base is e, we write $x = \log_e n$.

In this form **x is expressed as a function of n,** whereas in the form $n = 10^x$, n is expressed as a function of x.

The student must be able to change readily from one form to another.

Examples.

(1) We saw in § 143 that $56 \cdot 2 = 10^{1 \cdot 75}$.

In log form this is $1 \cdot 75 = \log_{10} 56 \cdot 2$.

(2) $$1024 = 2^{10}.$$

∴ $$\log_2 1024 = 10.$$

---

[1] The choice of the word logarithm can be explained only by the history of the word. The student could consult *A Short History of Mathematics*, by W. W. R. Ball.

(3) $$1000 = 10^3.$$
$$\therefore \quad \log_{10} 1000 = 3.$$
(4) $$81 = 3^4.$$
$$\therefore \quad \log_3 81 = 4.$$

For ordinary calculations 10 is the most suitable base for a system of logarithms, but in more advanced mathematics a different base is required (see § 153).

146. Characteristic of a logarithm.

The integral or whole number part of a logarithm is called the characteristic. This can always be determined by inspection when logarithms are calculated to base 10, as will be seen from the following considerations:

Since
$$10^0 = 1, \qquad \log_{10} 1 \quad = 0$$
$$10^1 = 10, \qquad \log_{10} 10 \quad = 1$$
$$10^2 = 100, \qquad \log_{10} 100 \quad = 2$$
$$10^3 = 1000, \qquad \log_{10} 1000 \quad = 3$$
$$10^4 = 10\,000, \log_{10} 10\,000 = 4$$

and so on.

From these results we see that,

for numbers between	1 and	10 the characteristic is 0
,, ,, ,,	10 ,, 100 ,,	,, ,, 1
,, ,, ,,	100 ,, 1000 ,,	,, ,, 2
,, ,, ,,	1000 ,, 10 000 ,,	,, ,, 3

and so on.

It is evident that **the characteristic is always one less than the number of digits in the whole number part of the number.**

Thus in
$$\log_{10} 3758 \cdot 7 \text{ the characteristic is } 3$$
$$\log_{10} 375 \cdot 87 \quad,, \qquad ,, \qquad ,, 2$$
$$\log_{10} 37 \cdot 587 \quad,, \qquad ,, \qquad ,, 1.$$

Thus the characteristics may always be determined by inspection, and consequently are not given in the tables. This is one advantage of having 10 for a base.

147. Mantissa of a logarithm.

The decimal part of a logarithm is called the mantissa.

In general the mantissa can be calculated to any required number of figures, by the use of higher mathematics. In most tables, such as those given in this volume, the mantissa is calculated to *four* places of decimals approximately. In *Chambers' " Book of Tables "* they are calculated to seven places of decimals.

The mantissa alone is given in the tables, and the following example will show the reason why:

$$\log_{10} 168 \cdot 3 = 2 \cdot 2261.$$
$$\therefore \qquad 168 \cdot 3 = 10^{2 \cdot 2261}.$$
$$\therefore \quad 168 \cdot 3 \div 10 = 10^{2 \cdot 2261} \div 10^1.$$
$$\therefore \qquad 16 \cdot 83 = 10^{2 \cdot 2261 - 1} \text{ (second law of indices)}$$
$$= 10^{1 \cdot 2261}.$$
$$\therefore \qquad \log_{10} 16 \cdot 83 = 1 \cdot 2261.$$
$$\text{Similarly} \quad \log_{10} 1 \cdot 683 = 0 \cdot 2261$$
$$\text{and} \qquad \log_{10} 1683 = 3 \cdot 2261.$$

Thus, if a number is multiplied or divided by a power of 10, the characteristic of the logarithm of the result is changed, but the mantissa remains unaltered. This may be expressed as follows:

Numbers having the same set of significant figures have the same mantissa in their logarithms.

148. To read a table of logarithms.

With the use of the above rules relating to the characteristic and mantissa of logarithms, the student should have no difficulty in reading a table of logarithms.

Below is a portion of such a table, giving the logarithms of numbers between 31 and 35.

No.	Log.	1	2	3	4	5	6	7	8	9	1	2	3	4	5	6	7	8	9
31	4914	4928	4942	4955	4969	4983	4997	5011	5024	5038	1	3	4	6	7	8	10	11	12
32	5051	5065	5079	5092	5105	5119	5132	5145	5159	5172	1	3	4	5	7	8	9	11	12
33	5185	5198	5211	5224	5237	5250	5263	5276	5289	5302	1	3	4	5	6	8	9	10	12
34	5315	5328	5340	5353	5366	5378	5391	5403	5416	5428	1	3	4	5	6	8	9	10	11
35	5441	5453	5465	5478	5490	5502	5514	5527	5539	5551	1	2	4	5	6	7	9	10	11

(1)　(2)

The figures in column 1 in the complete table are the numbers from 1 to 99. The corresponding number in column 2 is the mantissa of the logarithm. As previously stated, the characteristic is not given, but can be written down by inspection. Thus $\log_{10} 31 = 1.4914$, $\log_{10} 310 = 2.4914$, etc.

If the number has a *third significant figure*, the mantissa will be found in the appropriate column of the next nine columns.

Thus $\log_{10} 31.1 = 1.4928$,
                 $\log_{10} 31.2 = 1.4942$, and so on.

If the number has a *fourth significant figure* space does not allow us to print the whole of the mantissa. But the next nine columns of what are called " mean differences " give us for every fourth significant figure a number which must be added to the mantissa already found for the first three significant figures. Thus if we want $\log_{10} 31.67$, the mantissa for the first three significant figures 316 is 0.4997. For the fourth significant figure 7 we find in the appropriate column of mean differences the number 10. This is added to 0.4997 and so we obtain for the mantissa 5007.

$\therefore$ $\log_{10} 31.67 = 1.5007$.

Anti-logarithms.

The student is usually provided with a table of anti-logarithms which contains the *numbers corresponding to given logarithms*. These could be found from a table of logarithms but it is quicker and easier to use the anti-logarithms, which are given at the end of this book.

The tables are similar in their use to those for logarithms, but we must remember:

(1) That the mantissa of the log only is used in the table.

(2) When the significant figures of the number have been obtained, the student must proceed to fix the decimal point in them by using the rules which we have considered for the characteristic.

Example. *Find the number whose logarithm is* 2.3714.

First using the mantissa—viz., 0.3714—we find from the anti-logarithm table that the number corresponding is given as 2352. These are the first four significant figures of the number required.

Since the characteristic is 2, the number must lie between

100 and 1000 (see § 146) and therefore it must have 3 significant figures in the integral part.

∴ the number is 235·2.

Note.—As the log tables which will be usually employed by the beginner are all calculated to base 10, the base in further work will be omitted when writing down logarithms. Thus we shall write log 235·2 = 2·3714, the base 10 being understood.

Exercise 43.

1. Write down the characteristics of the logarithms of the following numbers:

15,   1500,   31 672,   ·597,   8,   800 000
        51·63,   3874·5,   2·615,   325·4

2. Read from the tables the logarithms of the following numbers:

(1) 5, 50, 500, 50 000.
(2) 4·7, 470, 47 000.
(3) 52·8, 5·28, 528.
(4) 947·8, 9·478, 94 780.
(5) 5·738, 96·42, 6972.

3. Find, from the tables, the numbers of which the following are the logarithms:

(1) 2·65, 4·65, 1·65.
(2) 1·943, 3·943, 0·943.
(3) 0·6734, 2·6734, 5·6734.
(4) 3·4196, 0·7184, 2·0568.

149. Rules for the use of logarithms.

In using logarithms for calculations we must be guided by the laws which govern operations with them. Since logarithms are indices, these laws must be the same in principle as those of indices. These laws are given below; formal proofs are omitted. They follow directly from the corresponding index laws.

(1) Logarithm of a product.

The logarithm of the product of two or more numbers is equal to the sum of the logarithms of these numbers (see first law of indices).

Thus if p and q be any numbers

$$\log (p \times q) = \log p + \log q.$$

(2) Logarithm of a quotient.

The logarithm of p divided by q is equal to the logarithm of p diminished by the logarithm of q (see second law of indices).

Thus $\qquad \log (p \div q) = \log p - \log q.$

(3) Logarithm of a power.

The logarithm of a power of a number is equal to the logarithm of the number multiplied by the index of the power (see third law of indices).

Thus $\qquad \log a^n = n \log a.$

(4) Logarithm of a root.

This is a special case of the above (3).

Thus
$$\log \sqrt[n]{a} = \log a^{\frac{1}{n}}$$
$$= \frac{1}{n} \log a.$$

150. Examples of the use of logarithms.

Example 1. *Find the value of* $57 \cdot 86 \times 4 \cdot 385.$

Let $\qquad x = 57 \cdot 86 \times 4 \cdot 385.$

Then $\quad \log x = \log 57 \cdot 86 + \log 4 \cdot 385$

No.	log.
57·86	1·7624
4·385	0·6420
253·7	2·4044

$$\begin{aligned} &= 1 \cdot 7624 + 0 \cdot 6420 \\ &= 2 \cdot 4044 \\ &= \log 253 \cdot 7. \end{aligned}$$

$\therefore \qquad x = 253 \cdot 7.$

Notes.—(1) The student should remember that the logs in the tables are correct to four significant figures only. Consequently he cannot be sure of four significant figures in the answer. It would be more correct to give the above answer as 254, correct to three significant figures.

(2) The student is advised to adopt some systematic way of arranging the actual operations with logarithms. Such a method is shown above.

Example 2. *Find the value of*

$$\frac{5\cdot672 \times 18\cdot94}{1\cdot758}.$$

Let $x = \dfrac{5\cdot672 \times 18\cdot94}{1\cdot758}.$

$\therefore \log x = \log 5\cdot672 + \log 18\cdot94 - \log 1\cdot758$
$= 0\cdot7538 + 1\cdot2774 - 0\cdot2450$
$= 1\cdot7862$
$= \log 61\cdot12.$

$\therefore \qquad x = 61\cdot12.$

or $\qquad x = 61\cdot1$ (to three significant figures).

No.	log.
5·672	0·7538
18·94	1·2774
	2·0312
1·758	0·2450
61·12	1·7862

Example 3. *Find the fifth root of* 721·8.

Let $\qquad x = \sqrt[5]{721\cdot8}$
$= (721\cdot8)^{\frac15}.$

Then $\qquad \log x = \frac15 \log 721\cdot8$ (see § 149 (4))
$= \frac15 (2\cdot8584)$
$= 0\cdot5717.$

$\therefore \qquad x = 3\cdot730.$

Exercise 44.

Use logarithms to find the values of the following:

1. $23\cdot4 \times 14\cdot73.$
2. $43\cdot97 \times 6\cdot284.$
3. $987\cdot4 \times 1\cdot415.$
4. $42\cdot7 \times 9\cdot746 \times 14\cdot36.$
5. $28\cdot63 \div 11\cdot95.$
6. $43\cdot97 \div 6\cdot284.$
7. $23\cdot4 \div 14\cdot73.$
8. $927\cdot8 \div 4\cdot165.$
9. $94\cdot76 \times 4\cdot195 \div 27\cdot94.$
10. $\dfrac{15\cdot36 \times 9\cdot47 \times 11\cdot48}{5\cdot632 \times 21\cdot85}.$
11. $(9\cdot478)^3.$
12. $(51\cdot47)^2.$
13. $(1\cdot257)^5.$

14. $(15\cdot23)^2 \times 3\cdot142.$
15. $(5\cdot98)^2 \div 16\cdot47.$
16. $\dfrac{(91\cdot5)^2}{4\cdot73 \times 16\cdot92}.$
17. $\dfrac{(8\cdot97)^2 \times (1\cdot059)^3}{57\cdot7}.$
18. $\dfrac{4798}{(56\cdot2)^2 \div (9\cdot814)^3}.$
19. $\sqrt[4]{3\cdot417}.$
20. $\sqrt[3]{4\cdot872}.$
21. $\sqrt[3]{1\cdot625^2 \times 4\cdot738}.$
22. $\sqrt[5]{61\cdot5 \times 2\cdot73}.$

23. If $\pi r^2 = 78\cdot6$ find r when $\pi = 3\cdot142.$
24. If $\frac43 \pi r^3 = 15\cdot5$, find r when $\pi = 3\cdot142.$

151. Logarithms of numbers between 0 and 1.

In § 146 we gave examples of powers of 10 when the index is a positive integer. We will now consider cases in which the indices are negative.

Thus $10^1 = 10$ $\therefore$ $\log_{10} 10$ =   1
       $10^0 = 1$ $\therefore$ $\log_{10} 1$ =   0
       $10^{-1} = \frac{1}{10} = 0\cdot1$ $\therefore$ $\log_{10} 0\cdot1$ = $-$ 1
       $10^{-2} = \frac{1}{10^2} = 0\cdot01$ $\therefore$ $\log_{10} 0\cdot01$ = $-$ 2
       $10^{-3} = \frac{1}{10^3} = 0\cdot001$ $\therefore$ $\log_{10} 0\cdot001$ = $-$ 3

                                     etc.

From these results we may deduce that:

The logarithms of numbers between 0 and 1 are always negative.

We have seen (§ 147) that if a number be divided by 10, we obtain the log of the result by subtracting 1.

Thus if $\log 49\cdot8$ $= 1\cdot6972$
            $\log 4\cdot98$ $= 0\cdot6972$
            $\log 0\cdot498$ $= 0\cdot6972 - 1$
            $\log 0\cdot0498$ $= 0\cdot6972 - 2$
            $\log 0\cdot004\ 98 = 0\cdot6972 - 3.$
From the above, $\log 0\cdot498 = 0\cdot6972 - 1$
                        $= - 0\cdot3028.$

Now, in the logs of numbers greater than unity, the mantissa remains the same when the numbers are multiplied or divided by powers of 10 (see § 147), *i.e.* with the same significant figures we have the same mantissa.

It would clearly be a great advantage if we could find a system which would enable us to use this rule for numbers less than unity, and so avoid, for example, having to write

$$\log 0\cdot498 \text{ as } - 0\cdot3028.$$

This can be done by not carrying out the subtraction as shown above, and writing down the characteristic as negative. But to write $\log 0\cdot498$ as $0\cdot6972 - 1$ would be awkward. Accordingly we adopt the notation $\bar{1}\cdot6972$, writing the minus sign above the characteristic.

It is very important to remember that

$$\bar{1}\!\cdot\!6972 = -1 + 0\!\cdot\!6972.$$

Thus in logarithms written in this way **the characteristic is negative and the mantissa is positive.**

With this notation   log 0·0498   = $\bar{2}\!\cdot\!6972$
                     log 0·004 98 = $\bar{3}\!\cdot\!6972$
                     log 0·000 498 = $\bar{4}\!\cdot\!6972$ etc.

Note.—The student should note that *the negative characteristic is numerically one more than the number of zeros after the decimal point.*

Example 1.   *From the tables find the logs of* 0·3185, 0·031 85 *and* 0·003 185.

Using the portion of the tables in § 148, we see that the mantissa for 0·3185 will be 0·5031.

Also the characteristic is − 1.

∴   log 0·3185   = $\bar{1}\!\cdot\!5031$.
Similarly   log 0·031 85 = $\bar{2}\!\cdot\!5031$
and   log 0·003 185 = $\bar{3}\!\cdot\!5031$.

Example 2.   *Find the number whose log is* $\bar{3}\!\cdot\!5416$.

From the anti-log tables we find that the significant figures of the number whose mantissa is 5416 are 3480. As the characteristic is − 3, there will be two zeros after the decimal point.

∴ the number is   0·003 480

(correct to 4 significant figures).

Exercise 45.

1. Write down the logarithms of:
    (1) 2·798, 0·2798, 0·027 98.
    (2) 4·264, 0·4264, 0·004 264.
    (3) 0·009 783, 0·000 978 3, 0·9783.
    (4) 0·064 51, 0·6451, 0·000 645 1.

2. Write down the logarithms of:
    (1) 0·059 86.   (4) 0·000 092 75.
    (2) 0·000 473.   (5) 0·5673.
    (3) 0·007 963.   (6) 0·079 86.

3. Find the numbers whose logarithms are:

(1) $\bar{1}\cdot3342$.

(2) $\bar{3}\cdot8724$.

(3) $\bar{2}\cdot4871$.

(4) $\bar{4}\cdot6437$.

(5) $\bar{1}\cdot7738$.

(6) $\bar{8}\cdot3948$.

---

152. Operations with logarithms which are negative.

Care is needed in operating with the logarithms of numbers which lie between 0 and 1, since they are negative and, as shown above, are written with the characteristic negative and the mantissa positive.

A few examples will show the method of working.

Example 1. *Find the sum of the logarithms :*

$$\bar{1}\cdot6173, \quad \bar{2}\cdot3415, \quad \bar{1}\cdot6493, \quad 0\cdot7374.$$

Arranging thus

$$\begin{array}{r} \bar{1}\cdot6173 \\ \bar{2}\cdot3415 \\ \bar{1}\cdot6493 \\ 0\cdot7374 \\ \hline \bar{2}\cdot3455 \end{array}$$

The point to be specially remembered is that the 2 which is carried forward from the addition of the mantissæ is positive, since they are positive. Consequently the addition of the characteristics becomes

$$-1-2-1+0+2 = -2.$$

Example 2. *From the logarithm* $\bar{1}\cdot6175$ *subtract the* log $\bar{3}\cdot8463$.

$$\begin{array}{r} \bar{1}\cdot6175 \\ \bar{3}\cdot8463 \\ \hline 1\cdot7712 \end{array}$$

Here in " borrowing " to subtract the 8 from the 6, the -1 in the top line becomes -2, consequently on subtracting the characteristics we have

$$-2-(-3) = -2+3 = +1.$$

Example 3. *Multiply* $\bar{2}\cdot8763$ *by* 3.

$$\bar{2}\cdot8763$$
$$3$$
$$\overline{\bar{4}\cdot6289}$$

From the multiplication of the mantissa, 2 is carried forward. But this is positive and as $(-2) \times 3 = -6$, the characteristic becomes $-6 + 2 = -4$.

Example 4. *Multiply* $\bar{1}\cdot8738$ *by* $1\cdot3$.

In a case of this kind it is better to multiply the characteristic and mantissa separately and add the results.

Thus
$$0\cdot8738 \times 1\cdot3 = 1\cdot135\ 94$$
$$-1 \times 1\cdot3 = -1\cdot3.$$

$-1\cdot3$ is wholly negative and so we change it to $\bar{2}\cdot7$, to make the mantissa positive.

Then the product is the sum of

$$1\cdot135\ 94$$
$$\bar{2}\cdot7$$
$$\overline{\bar{1}\cdot835\ 94}$$

or $\quad\quad 1\cdot8359$ approx.

Example 5. *Divide* $\bar{5}\cdot3716$ *by* 3.

Here the difficulty is that on dividing $\bar{5}$ by 3 there is a remainder 2 which is negative, and cannot therefore be carried on to the positive mantissa. To get over the difficulty we write:

$$-5 = -6 + 1$$

or the log as $\quad -6 + 1\cdot3716.$

Then the division of the -6 gives us -2 and the division of the positive part $1\cdot3716$ gives $0\cdot4572$, which is positive. Thus the complete quotient is $\bar{2}\cdot4572$. The work might be arranged thus:

$$3)\overline{\bar{6} + 1\cdot3716}$$
$$\overline{\bar{2} + 0\cdot4572}$$
$$= \bar{2}\cdot4572$$

Exercise 46.

1. Add together the following logarithms:
    (1) $\bar{2}\cdot5178 + 1\cdot9438 + 0\cdot6138 + \bar{5}\cdot5283.$
    (2) $3\cdot2165 + \bar{3}\cdot5189 + \bar{1}\cdot3297 + \bar{2}\cdot6475.$

2. Find the values of:
    (1) $4\cdot2183 - 5\cdot6257.$ (3) $\bar{1}\cdot6472 - \bar{1}\cdot9875.$
    (2) $0\cdot3987 - \bar{1}\cdot5724.$ (4) $\bar{2}\cdot1085 - \bar{5}\cdot6271.$

3. Find the values of:
    (1) $\bar{1}\cdot8732 \times 2.$ (4) $\bar{1}\cdot5782 \times 1\cdot5.$
    (2) $\bar{2}\cdot9456 \times 3.$ (5) $\bar{2}\cdot9947 \times 0\cdot8.$
    (3) $\bar{1}\cdot5782 \times 5.$ (6) $\bar{2}\cdot7165 \times 2\cdot5.$

4. Find the values of:
    (1) $\bar{3}\cdot9778 \times 0\cdot65.$ (4) $2\cdot1342 \times -0\cdot4.$
    (2) $\bar{2}\cdot8947 \times 0\cdot84.$ (5) $1\cdot3164 \times -1\cdot5.$
    (3) $\bar{1}\cdot6257 \times 0\cdot6.$ (6) $\bar{1}\cdot2976 \times -0\cdot8.$

5. Find the values of:
    (1) $\bar{1}\cdot4798 \div 2.$ (4) $\bar{3}\cdot1195 \div 2.$
    (2) $\bar{2}\cdot5637 \div 5.$ (5) $\bar{1}\cdot6173 \div 1\cdot4.$
    (3) $\bar{4}\cdot3178 \div 3.$ (6) $\bar{2}\cdot3178 \div 0\cdot8.$

---

153. Change of base of a system of logarithms.

Although logs calculated to base 10 are usually employed for calculations, in more advanced Mathematics, as well as in Engineering, the logs which naturally arise are calculated to a base which is given by the series

$$1 + \frac{1}{1} + \frac{1}{1\,.\,2} + \frac{1}{1\,.\,2\,.\,3} + \frac{1}{1\,.\,2\,.\,3\,.\,4} + \ldots \text{ to infinity.}$$

This series is denoted by **e**, and its value can be calculated to any required degree of accuracy by taking sufficient terms. To 5 places of decimals, e = 2·718 28.

Logs calculated to this base are called **Naperian logarithms**, after Lord Napier, who discovered them in 1614, using this base. They are also called **Natural logarithms** or **Hyperbolic logarithms**.

The student, possessing only tables of logs to base 10 may require to use the logs of numbers to base e, and must therefore know how to find them.

The relations between the logs of numbers to different bases is found as follows:

Let n be any number.
Let a and b be two bases.

Suppose that logs to base b are known, and we require to find them to base a.

Let $\quad \log_b n = y$, $\therefore n = b^y$.
Then $\quad \log_a n = \log_a (b^y)$
$\qquad\qquad = y \log_a b$.   (§ 149, rule 3.)
$\therefore \qquad \log_a n = \log_b n \times \log_a b$.

Thus, knowing the log of a number to a base b, we find its log to base a by multiplying, whatever the number, by $\log_a b$.

In the above result let $b = 10$ and $a = e$.

Then $\qquad\qquad \log_e n = \log_{10} n \times \log_e 10$.

In this result $\qquad$ let $n = e$
then $\qquad\qquad\qquad \log_e e = \log_{10} e \times \log_e 10$
$\qquad\qquad$ but $\log_e e = 1$.
$\therefore \qquad\quad \log_{10} e \times \log_e 10 = 1$.

$\therefore \qquad\qquad\quad \log_e 10 = \dfrac{1}{\log_{10} e}$.

$\therefore$ in the rule $\quad \log_e n = \log_{10} n \times \log_e 10$
we can write $\quad \log_e n = \log_{10} n \times \dfrac{1}{\log_{10} e}$.

Thus both logs on the right-hand side are to base 10.

Now $\qquad\qquad \log_{10} e = 0 \cdot 4343$
and $\qquad\qquad \log_e 10 = \dfrac{1}{0 \cdot 4343} = 2 \cdot 3026$.

Hence to change from base 10 to base e, we may use either of the following:

$\qquad$ (1) $\log_e n = \log_{10} n \times 2 \cdot 3026$
or $\qquad$ (2) $\log_e n = \log_{10} n \div 0 \cdot 4343$.

Example. *Find* $\log_e 50$.

Using $\qquad\qquad \log_e 50 = \log_{10} 50 \times 2{\cdot}3026$
we have $\qquad\quad \log_e 50 = 1{\cdot}6990 \times 2{\cdot}3026$.

Evaluating the right-hand side by use of logs we get

$$\log_e 50 = 3{\cdot}913.$$

Summary of the laws relating to logarithms, together with some special points the truth of which will be obvious.

(1) The logarithm of the base itself is always unity.

(2) The logarithm of 1 is always zero, whatever the base.

(3) The logarithms of all numbers less than unity are negative.

(4) The logarithm of a number is equal to — (the log of its reciprocal).

Thus $\qquad\qquad\qquad \log_a n = -\log_a \dfrac{1}{n}$.

(5) $\qquad\quad \log_a (x \times y) = \log_a x + \log_a y$.

(6) $\qquad\quad \log_a (x \div y) = \log_a x - \log_a y$.

(7) $\qquad\qquad \log_a x^n = n \log_a x$.

(8) $\qquad\qquad \log_a \sqrt[n]{x} = \dfrac{1}{n} \log_a x$.

Exercise 47.

Miscellaneous Exercises in the Use of Logarithms.

1. $15{\cdot}62 \times 0{\cdot}987$.
2. $0{\cdot}4732 \times 0{\cdot}694$.
3. $0{\cdot}513 \times 0{\cdot}0298$.
4. $75{\cdot}94 \times 0{\cdot}0916 \times 0{\cdot}8194$.
5. $9{\cdot}463 \div 15{\cdot}47$.
6. $0{\cdot}9635 \div 29{\cdot}74$.
7. $27{\cdot}91 \div 569{\cdot}4$.
8. $0{\cdot}0917 \div 0{\cdot}5732$.
9. $5{\cdot}672 \times 14{\cdot}83 \div 0{\cdot}9873$.
10. $(0{\cdot}9173)^2$.
11. $(0{\cdot}4967)^3$.
12. $\sqrt[3]{1{\cdot}715}$.
13. $\sqrt[5]{647{\cdot}2} \div (3{\cdot}715)^3$.
14. $\tfrac{1}{2}(48{\cdot}62)^{\frac{1}{2}}$.
15. $\sqrt[3]{\dfrac{9{\cdot}728}{3{\cdot}142}}$.
16. $(1{\cdot}697)^{2{\cdot}4}$.
17. $(19{\cdot}72)^{0{\cdot}57}$.
18. $(0{\cdot}478)^{3{\cdot}1}$.
19. $(5{\cdot}684)^{-1{\cdot}12}$.
20. $(0{\cdot}5173)^{-3{\cdot}4}$.
21. $\sqrt[4]{0{\cdot}016\ 97}$.
22. $(0{\cdot}1478)^2 \div 0{\cdot}6982$.
23. $\sqrt[3]{0{\cdot}8172} \div \sqrt[3]{0{\cdot}7658}$.

24. $9.74^2 - 5.66^2$. (Hint.—This should first be factorized and changed to a product.)

25. $\dfrac{9.32 \times 0.761}{\sqrt{18.2}}$.

26. $\sqrt[3]{\dfrac{647.3 \times 3.2}{3.142 \times 10.78}}$.

27. $\sqrt{(3.62)^2 + (5.47)^2 + (6.91)^2}$.

28. Find the value of πr^2 when $\pi = 3.142$ and $r = 16.89$.

29. Find the value of $\frac{4}{3}\pi r^3$ when $\pi = 3.142$ and $r = 2.9$.

30. If $V = pv^{1.6}$ find V when $v = 6.032$ and $p = 29.12$.

31. If $3^x = 24$, find x.

32. If $R^n = 1.8575$ find R when $n = 18$.

33. When $l = \dfrac{gt^2}{\pi^2}$ find l when $g = 9.81, t = 2, \pi = 3.142$.

34. From the formula $V = \sqrt{\dfrac{2ghD}{0.03L}}$ find V when $g = 9.8$, $h = 0.627$, $L = 175$, $D = 0.27$.

35. If $t = 2\pi\sqrt{\dfrac{l}{g}}$, find g when $l = 159$ mm, $t = 0.8$ s, $\pi = 3.142$.

36. Without using tables find the values of:

      (a) $\log 27 \div \log 3$.   (b) $(\log 16 - \log 2) \div \log 2$.

37. If $M = PR^n$ find M when $P = 200$, $R = 1.05$, $n = 20$.

38. Find the radius of a sphere whose volume is 500 mm³. $V = \frac{4}{3}\pi r^3$.

39. Find the values of (1) $\log_e 4.6$

                  (2) $\log_e 0.062$.

40. The insulating resistance, R, of a wire of length l is given by

$$R = \frac{0.42S}{l} \times \log_e \frac{d_2}{d_1}.$$

Find l when $S = 2000$, $R = 0.44$, $d_2 = 0.3$, $d_1 = 0.16$.

41. In a calculation on the dryness of steam the following formula was used:

$$\frac{qL}{T} = \frac{q_1 L_1}{T_1} + \log_e \frac{T_1}{T}.$$

Find q when $L_1 = 850$, $L = 1000$, $T_1 = 780$, $T = 650$, $q_1 = 1$.

CHAPTER XVII

RATIO AND PROPORTION

154. Meaning of a ratio.

THERE are two ways of comparing the magnitudes of two numbers:

> (1) **By subtraction.** This operation states by how much one number is greater or less than the other. If the numbers are represented by a and b, the comparison is expressed by $a - b$.
>
> (2) **By division.** By this means we learn what multiple or what part or parts one number is of the other.

The latter is called the **ratio** of the two numbers, and may be expressed by $a \div b$, or $\frac{a}{b}$, or in the special form $a : b$. Of these, the fractional form is best suited for manipulation.

155. Ratio of two quantities.

The magnitude of **two quantities of the same kind,** such as two lengths, weights, sums of money, etc., may be compared by means of a ratio. To do this the measures of the two quantities are expressed in terms of the **same unit** by numbers. The ratio of these two numbers expresses the ratio of the quantities.

Thus, the ratio of two distances which are respectively a metres and b metres would be $\frac{a}{b}$ or $a : b$.

The ratio would be unaltered in value if the quantities were expressed in other units.

Thus the ratio of 3 hours to 2 hours is the same as the ratio of 9 metres to 6 metres or 108° to 72°.

This is obvious from the consideration that, as a ratio can

be expressed by a fraction, it can be manipulated in the same ways as fractions.

Thus
$$\frac{a}{b} = \frac{a \times m}{b \times m}$$

and
$$\frac{a}{b} = \frac{a \div m}{b \div m}.$$

A ratio is always a number, either an integer or a fraction (vulgar or decimal), and is not expressed in terms of any particular unit.

156. Proportion.

If four numbers a, b, c, d are so related that the ratios $\frac{a}{b}$ and $\frac{c}{d}$ are equal, the numbers are said to be in proportion.

It follows from the definition of a ratio that a and b must represent the ratio of two quantities of the **same** kind, while c and d must also represent the ratio of two quantities of the same kind, though not necessarily of the same kind as a and b.

Thus a and b might represent the measures of two **weights**, while c and d ,, ,, ,, **costs**.

Continued proportion.

If a series of numbers a, b, c, d . . . is such that

$$\frac{a}{b} = \frac{b}{c} = \frac{c}{d} = \cdots$$

then these numbers are said to be in continued proportion.

Thus in the series of numbers 2, 6, 18, 54 . . . the ratios

$$\frac{2}{6} = \frac{6}{18} = \frac{18}{54} \cdots$$

are all equal. The numbers are in continued proportion.

Mean proportional. If a, b, c are numbers such that

$$\frac{a}{b} = \frac{b}{c},$$

b is called a **mean proportional** between a and c.

Then
$$b^2 = ac$$
and
$$b = \sqrt{ac}.$$

In this way we can find a mean proportional between any two numbers.

157. Theorems on ratio and proportion.

The following theorems should be noted:

(1) Let
$$\frac{a}{b} = \frac{c}{d}.$$

Then
$$\frac{a}{b} \times bd = \frac{c}{d} \times bd.$$

$$\therefore \quad ad = bc.$$

(2) Let
$$\frac{a}{b} = \frac{c}{d}.$$

Then from the meaning of a ratio
$$\frac{b}{a} = \frac{d}{c} \quad (alternando).$$

(3) Let
$$\frac{a}{b} = \frac{c}{d}.$$

Then
$$ad = bc \quad (\text{theorem 1}).$$

$$\therefore \quad \frac{ad}{cd} = \frac{bc}{cd}.$$

$$\therefore \quad \frac{a}{c} = \frac{b}{d} \quad (invertendo).$$

(4) Let
$$\frac{a}{b} = \frac{c}{d}.$$

Then
$$\frac{a}{b} + 1 = \frac{c}{d} + 1.$$

$$\therefore \quad \frac{a+b}{b} = \frac{c+d}{d} \quad (componendo).$$

(5) Let
$$\frac{a}{b} = \frac{c}{d}.$$

Then
$$\frac{a}{b} - 1 = \frac{c}{d} - 1.$$

$$\therefore \quad \frac{a-b}{b} = \frac{c-d}{d} \quad (dividendo).$$

(6) By division of (4) by (5)
$$\frac{a+b}{a-b} = \frac{c+d}{c-d}.$$

This is called *componendo and dividendo*.

158. An Illustration from Geometry.

The following illustration from Geometry is given as being of great importance. For a proof of it and for a geometrical treatment of ratio and proportion the student is referred to any text-book on Geometry.

Similar triangles. Triangles which have their corresponding angles equal are called **similar**. In such triangles,

the ratios of corresponding sides are equal.

In Fig. 42 the triangles ABC, $A'B'C'$ are similar—*i.e.*,

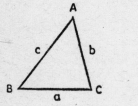

Fig. 42.

$\angle A = \angle A'$, $\angle B = \angle B'$, $\angle C = \angle C'$. Corresponding sides are those which are opposite to equal angles.

Denoting the lengths of the sides opposite to these by a, b, c, a', b', c' as shown, then by the theorem stated above

$$\frac{a}{a'} = \frac{b}{b'} = \frac{c}{c'}.$$

It follows from theorem (3) of § 157

$$\frac{a}{b} = \frac{a'}{b'}, \ \frac{b}{c} = \frac{b'}{c'} \text{ and } \frac{a}{c} = \frac{a'}{c'},$$

i.e., the ratios of the pairs of sides containing the equal angles are equal. Each pair of equal ratios gives a set of numbers in proportion.

159. Constant ratios.

Let AOB be any angle (Fig. 43).

On the arm OA take points P, Q, R . . . and draw PK, QL, RM perpendicular to the other arm.

Then the $\triangle$s OPK, OQL, ORM are equiangular and similar.

∴ by the theorem of § 158

$$\frac{PK}{OK} = \frac{QL}{OL} = \frac{RM}{OM}$$

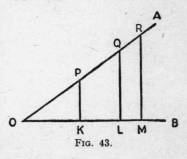

Clearly any number of points, such as P, Q, R, can be taken and the ratio of all pairs of sides such as the above are equal. We can therefore say that for the angle AOB all such ratios are constant in value.

This constant ratio is called in Trigonometry the tangent of the angle

FIG. 43.

AOB. It is abbreviated to tan AOB—i.e.,

$$\frac{PK}{OK} = \frac{QL}{OL} = \frac{RM}{OM} = \tan AOB.$$

A similar result is true for any other angle.

∴ every angle has its own tangent or constant ratio, by which it can be identified.

Referring to §§ 72 and 73 it will be seen that the gradient of a straight line, represented by m, is the tangent of the angle made by the line with the x axis.

In Fig. 24, for example, the ratio $\frac{PQ}{OQ}$ is constant for every point on the line, and is the tangent of the angle POQ.

In the general equation $y = mx + b$, m represents the tangent of the angle made with the axis of x.

160. Examples of equal ratios.

Examples of equal ratios frequently occur in Mathematics, and the following theorem, in different forms, is sometimes useful.

Let $\frac{a}{b} = \frac{c}{d} = \frac{e}{f}$ be equal ratios (there may be any number of them).

Let k represent their common value.

Then
$$\frac{a}{b} = \frac{c}{d} = \frac{e}{f} = k.$$

$\therefore$
$$a = bk$$
$$c = dk$$
$$e = fk.$$

These results make possible various manipulations. For example, by addition

$$a + c + e = bk + dk + fk$$
$$= k(b + d + f).$$

$\therefore$
$$\frac{a + c + e}{b + d + f} = k,$$

and is therefore equal to each of the original ratio. This can be varied in many ways.

For example, by multiplication of the three equations above:

$$2a = 2bk$$
$$3c = 3dk$$
$$7e = 7fk.$$

$\therefore$ by addition

$$2a + 3c - 7e = 2bk + 3ak - 7fk$$
$$= k(2b + 3d - 7f).$$

$\therefore$
$$\frac{2a + 3c - 7e}{2b + 3d - 7f} = k = \text{each of the original ratio.}$$

Exercise 48.

1. Write down the following ratios:

(a) £a to b pence.

(b) p h to q min r s.

2. Write down the ratio of " a km an hour " to " a m per second ".

3. If the ratio $a : b$ is equal to the ratio $5 : 8$, find the numerical values of the following ratios:

(1) $\dfrac{1}{a} : \dfrac{1}{b}$.   (2) $a^2 : b^2$.   (3) $2a : 3b$.

4. (a) Two numbers are in the ratio of $4 : 5$. If the first is 28, what is the second?

    (b) Two numbers are in the ratio of $a : b$. If the first is x, what is the second?

    (c) Two numbers are in the ratio of $1 : x$. If the second is a, what is the first?

5. If the ratio $\dfrac{a}{b+c}$ is equal to the ratio $\dfrac{c}{x}$, find x.

6. A rectangle of area A mm² is divided into two parts in the ratio $p : q$. Find expressions for their areas.

7. A piece of metal of mass a kg is divided into two parts in the ratio $x : y$. What are the masses of the parts?

8. If $\dfrac{1}{a}, \dfrac{1}{b}, \dfrac{1}{c}, \dfrac{1}{x}$ are four numbers in proportion, find x.

9. What number must be added to each term of the ratio $11 : 15$, so that it becomes the ratio $7 : 8$?

10. Find the mean proportional between:

    (1) ab and bc.   (2) $8a^2$ and $2b^2$.
    (3) $a(a + b)$ and $b(a + b)$.

11. What number added to each of the numbers 2, 5, 8, 13 will give four numbers in proportion?

12. If $\dfrac{a}{b} = \frac{5}{2}$, find the value of $\dfrac{a+b}{a-b}$.

13. Find the ratio of $\dfrac{a}{b}$ when

    (1) $3a = 7b$.   (2) $\dfrac{3}{a} = \dfrac{7}{b}$.
    (3) $16a^2 = 25b^2$.

14. (1) What is the result when 270 is altered in the ratio $5 : 3$?

    (2) What is the result of altering $\dfrac{a}{b}$ in the ratio $6 : 5$?

    (3) What is the result of altering $\frac{4}{3}$ in the ratio $a : b$?

15. If a, x, y, b are in continued proportion, find x and y in terms of a and b.

CHAPTER XVIII

VARIATION

161. Direct variation.

In Chapter XIII examples are given of a variable quantity the value of which depended on the value of another variable, and is called a function of it. There are very many different forms which this variation may take, and in this chapter we shall examine one of the most important of them. We will begin with a very simple example.

If a man is paid at a certain rate for every hour that he works, the total amount he earns over any period depends on the number of hours he works. If he doubles the number of hours he works, his earnings will be doubled. Generally the ratio of the amounts he earns in any two periods is equal to the ratio of the number of hours worked in the periods.

If T_1 and T_2 represent the number of hours worked in two periods and if W_1 and W_2 represent the amounts of wages earned in them,

then
$$\frac{W_1}{W_2} = \frac{T_1}{T_2}.$$

These four quantities are in proportion (see § 156).

Hence, when the relations between the two quantities can be expressed in this way we say that

> **The wages are proportional to the time worked or the wages vary directly as the time.**

The wages are a function of the time, and the words " proportional to " or " vary directly as " are used to define the exact functional relations which exist between the two quantities.

Direct variation may be defined as follows:

If two quantities Y and X are so related that $\dfrac{Y_1}{X_1} = \dfrac{Y_2}{X_2}$, *where X_1 and X_2 are any two values of X and Y_1 and Y_2 are corresponding values of Y, then Y is said to be proportional to X, or Y varies directly as X.*

In order to discover whether or not one quantity varies directly as another, this simple test can be applied.

If one quantity is doubled, is the other doubled in consequence? Or, more precisely, *if one quantity is altered in a certain ratio, is the other altered in the same ratio?*

The sign $\propto$ is used to denote direct variation. Thus in the case above we would write $Y \propto X$.

162. Examples of direct variation.

(1) The **distance** travelled by a car moving with uniform speed varies directly as the **time**.

(2) The **mass** of an amount of water is proportional to the **volume**.

(3) The **circumference** of a circle varies directly as the **diameter**.

(4) The **electrical resistance** of a wire varies directly as the **length**.

163. The constant of variation.

If $y \propto x$ then $y = kx$, where k is a constant.

In § 160 the common value of a number of equal ratios was represented by k, a **constant** number for these ratios.

When one quantity varies directly as another, then we have seen that the ratio of corresponding pairs of values of the variables is constant. Consequently, as in § 160, this constant is usually represented by k, and by means of it the relation between the quantities can be expressed as a formula.

For instance, in Example (1) of § 162 it was stated that for a body moving with uniform speed, **distance** moved varies directly as **time**.

Let s represent any distance travelled.

,, t ,, the corresponding time.

Then, since s varies directly as t, any ratio of corresponding values of these is constant.

Let k represent this constant.

Then by definition $\qquad \dfrac{s}{t} = k.$

$\therefore \qquad\qquad\qquad\quad s = kt.$

This is the form of the law connecting s and t, but it cannot be of much use until the value of k is known for this particular case.

To find this, we must know a pair of corresponding values of s and t. Thus if we are told that the car moves 40 m in 2·5 s, then, substituting the values for s and t, we get

$$40 = k \times 2·5.$$
$$\therefore \qquad k = 40 \div 2·5$$
$$= 16.$$

$\therefore$ the law connecting s and t for this particular velocity is
$$s = 16t.$$

164. Graphical representation.

If x and y represent two variables such that $y \propto x$, then, as shown in § 163, $y = kx$.

The form of this equation is the same as $y = mx$, the graph of which was shown in § 72 to be a straight line passing through the origin, m being the gradient of the line.

$\therefore y = kx$ represents a straight line of gradient equal to k.

Consequently the graphical representation of the variation of two quantities where one varies directly as the other is a straight line passing through the origin.

165. To find the law connecting two variables.

The engineer and the scientist frequently require to know the law connecting two variables, corresponding values of which have been found by experiments. The law may assume many forms, but in some cases there may be reason to suppose that one of the quantities varies directly as the other—i.e., the law may be of the form $y = kx$.

From the results of the experiment we can proceed to determine:

(1) Is the law one of direct variation?
(2) If it is, and the law is of the form $y = kx$, the value of k must be found.

To determine (1) the results are plotted. Then:

(a) If the graph is a straight line, the law is one of direct variation.

(b) If the straight line passes through the origin, the equation is of the form $y = kx$.

Then k has to be determined.

Graphically, k can be found as in § 159 by finding the tangent of the angle made with the axis of x.

Algebraically as shown in § 163. A pair of corresponding values of x and y is chosen from the graph. These are substituted in $y = kx$, and so k is determined.

166. Worked example.

A spiral spring is extended by hanging various weights from it.

The amounts of extension of the spring for different weights were observed and tabulated as follows:

Weight in N .	0	0·1	0·2	0·3	0·4	0·5
Extension in cm .	0	0·15	0·3	0·44	0·6	0·75

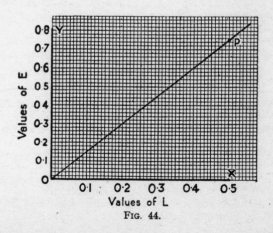

Fig. 44.

From these results discover the law connecting the attached weight and the extension of the spring.

The graph resulting from plotting these results is shown in Fig. 44. It is a straight line passing through the origin, though one of the points, corresponding to a weight of 0·3 N, lies slightly off the line. This is to be expected in experimental results. Also the line passes through the origin.

∴ as shown in § 165, the law connecting the weight and extension is one of direct variation—*i.e.*,

The extension varies directly as the attached load.

Let E = extension

  ,, L = load

then $E \propto L$ and $E = kL$.

To find k a pair of values is taken, at the point P where

$$E = 0·75 \text{ and } L = 0·5.$$

Substituting these $0·75 = k \times 0·5$

whence $k = 1·5$.

∴ the law is $E = 1·5L$.

167. y partly constant and partly varying as x.

The case frequently occurs, in practical work, of a variable quantity which in part varies directly as another quantity and in part is constant.

There is an example of this in § 67 in the problem concerning the profits of a restaurant. The profit depends on

    (1) The number of customers which is **variable**.

    (2) The overhead charges which are **constant**.

It was found that $y = ax - b$ was the law which connected these, where y, the profit, varies directly as x, the number of customers, and also depends on the constant charges b.

In general all such cases can be represented by the equation $y = kx + b$.

This is the equation of a straight line, but **it does not pass through the origin** (§ 73). It contains the two constants k and b, which must be determined before the law connecting x and y can be stated.

If two pairs of values of x and y are known, the solution can be found as shown in § 59, example 2.

In practical work pairs of values are found by experiment. The worked example which follows shows the method of procedure in such cases.

168. Worked example.

When two voltmeters are compared they have readings corresponding to C and K below.

C .	1·9	2·75	3·8	4·8	5·8
K .	5·75	8·3	11·2	14	16·8

It is thought that C and K are connected by a law of the form

$$K = mC + b.$$

Test this by plotting the points and find the values of m and b.

The law $K = mC + b$ is linear—*i.e.*, it is in the general form of the equation of a straight line. To test them we must find if the plotted points be on a straight line.

Comparison of the two sets of values suggests that the

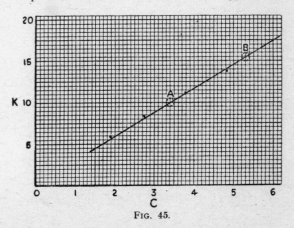

FIG. 45.

scale units for C on the axis of x should be larger than those for K on the y axis.

When the points are plotted they are seen to lie approximately on a straight line, slight deviations being due to experimental errors. When this line is drawn as evenly as possible, it appears as in Fig. 45.

Two suitable points, A and B, are selected on the line, and their co-ordinates are as follows:

For A, $C = 3.4$, $K = 10$.

" B, $C = 5.3$, $K = 15.5$.

These are to satisfy the equation
$$K = mC + b.$$

Substituting we get:
$$10 = 3.4m + b \quad . \quad . \quad . \quad . \quad . \quad (1)$$
$$15.5 = 5.3m + b \quad . \quad . \quad . \quad . \quad . \quad (2)$$

Subtracting,
$$1.9m = 5.5$$
$$\therefore \qquad m = \frac{5.5}{1.9} = 2.9 \text{ approx.}$$

Substituting for m in (1),
$$10 = (3.4 \times 2.9) + b.$$
$$\therefore \qquad b = 10 - 9.86$$
$$= 0.14.$$

$\therefore$ the law is $K = 2.9C + 0.14$.

Exercise 49.

1. The following are examples in which the value of one quantity depends on another. State in each case whether or not it is a case of direct variation:

   (a) *Distance* and *time* when a man runs the 1000 m race.

   (b) *Interest* and *time* when money bears interest at a fixed rate.

   (c) The *logarithm* of a number and the *number* itself.

   (d) The *y* co-ordinate of a point on a straight line and the *x* co-ordinate.

   (e) The *cost* of running a school and the *number* of scholars.

2. If $y = kx$ and $y = 8$ when $x = 7$, find k. Hence find y when $x = 40$.

3. If y is proportional to x and $y = 10$ when $x = 4$, find y when $x = 15$; also find x when $y = 8.4$.

4. If $y \propto x$, and when $y = 16.5$, $x = 3.5$, find the law connecting x and y. Hence find x when $y = 21$.

5. The distances travelled by a body from rest were as follows:

Time (s) .	1	2	3	4
Distance (m)	3·2	6·4	9·6	12·8

Plot these and find if distance varies directly as time. If it does, find:

    (1) The law connecting time (t) and distance (s).
    (2) The distance passed over in 2·8 s.

6. The extension (E) of a spring varies directly as the force (W) by which it is stretched. A certain spring extended 2·4 when stretched by a weight of 4·5. Find:

    (1) The law which connects them.
    (2) The extension due to a weight of 7.

7. The law connecting two variables x and y is of the form $y = kx + b$, where k and b are constants. When $x = 10$, $y = 11$, and when $x = 18$, $y = 15$. Find k and b and state the law.

8. In a certain machine the law connecting the applied force (E) and the load (W) was of the form $E = aW + b$, where a and b are constants. When $w = 20$, $E = 1.4$, and when $W = 30$, $E = 2$. Find the law.

9. In experiments to determine the friction (F N) between two metallic surfaces when the load is W N, the following results were found:

W .	3	5	7	10	12
F .	0·62	1·5	2·4	3·6	4·4

Assuming W and F to be connected by a law of the form $F = aW + b$, find the law by drawing the average straight line between the points.

Other forms of direct variation

169. y varies as the square of x—i.e., $y \propto x^2$.

If the sides of a square be doubled, the area is not doubled, but is **four** times as great; if the sides be trebled the area is **nine** times as great. **The area of a square varies directly as the square of the length of its sides.**

The ratio of the area of a circle to the square of the length of its radius can be shown experimentally to be constant. If $A =$ the area and $r =$ radius, then $\dfrac{A}{r^2}$ is constant for all circles. This constant is represented by the special symbol π, and its value is approximately 3·1416.

∴ **the area of a circle varies directly as the square of the length of its radius.**

Students of Mechanics will know that **the distance passed over by a body moving with uniformly increasing velocity is proportional to the square of the time.** A special case is a falling body.

If $s =$ the distance fallen and $t =$ the time taken in s, then $s \propto t^2$ and $s = kt^2$.

Experiments show that $k = \frac{1}{2}g$, where g is an absolute constant whose value is approximately 9·81 m/s².

The graphical representation of $y = kx^2$ is that of a quadratic function (see Chapter XIII). For different values of k, the graph is a parabola, symmetrical about the y axis and having its lowest point at the origin (§ 111).

170. y varies as the cube of x—i.e., $y \propto x^3$.

If the edge of a *cube* be doubled, the volume is 8 times as great. **The volume varies directly as the cube of the length of an edge.**

The volume of a *sphere* also varies as the cube of the radius.

If $V =$ the volume and $r =$ the length of the radius, then $V \propto r^3$ and $V = kr^3$.

It can be shown that $k = \frac{4}{3}\pi$.

$$\therefore \qquad V = \frac{4}{3}\pi r^3.$$

The *graph* of $y = x^3$ is a curve, as shown in Fig. 46. It is called a cubical parabola.

The student should make a table of values and draw the curve.

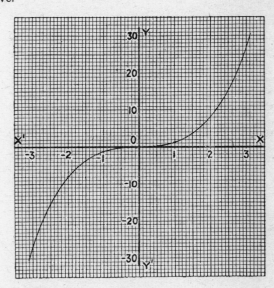

Fig. 46.

171. y varies as $\sqrt{x}$ or $x^{\frac{1}{2}}$, *i.e.*, $y \propto \sqrt{x}$.

This form of variation, besides arising in various physical examples, may also be regarded as the *inverse* of $y = x^2$.

Since
$$y = kx^2$$
$$x^2 = \frac{1}{k}y.$$

$$\therefore \qquad x = \sqrt{\frac{1}{k}} \times \sqrt{y}.$$

Since $\sqrt{\dfrac{1}{k}}$ is a constant, then $x \propto \sqrt{y}$.

The graph of $y = \sqrt{x}$ is a parabola, as in Fig. 47. It is the same curve as Fig. 31, as will be realised from §§ 108 and 109, but being the inverse, it is *symmetrical about the x axis*. Negative roots will be found on the part of the curve below the x axis.

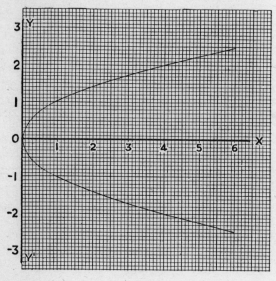

Fig. 47.

172. Inverse variation: $y \propto \dfrac{1}{x}$.

Let x and y be two numbers such that their product is constant—*i.e.*, $xy = k$,

then
$$y = \frac{k}{x} \text{ or } x = \frac{k}{y}.$$

Each quantity can be expressed in terms of the reciprocal or inverse of the other.

Many examples of this occur in Mathematics. The following is a simple case.

Let x and y be the sides of a rectangle of area 60 m², then $xy = 60$.

The lengths of the sides may be varied in very many ways, their product always being equal to 60. If x is increased, y will be decreased, and vice versa. If x be doubled, y will be halved. In general, if x be changed in a given ratio, y will be changed in the *inverse ratio*.

Hence we say that **y varies inversely as x**—*i.e.*, $y \propto \dfrac{1}{x}$.

Hence $$y = \frac{k}{x}.$$

Among the many examples of inverse variation we may note:

(1) *Time* to travel a given distance varies inversely as the speed.

If the speed be doubled, the time is halved.

(2) The *volume* of a fixed mass of gas varies inversely as the *pressure* on it, the temperature remaining constant.

If p = the pressure and v = the volume.

$$p \propto \frac{1}{v} \text{ or } p = \frac{k}{v}.$$

(3) The *electrical resistance* of a wire of given length and material to the passage of a current through it varies inversely as the *area of the cross-section of the wire*.

If R = the resistance,
and A = the area of the cross section,

then $$R \propto \frac{1}{A} \text{ or } R = \frac{k}{A}.$$

173. Graph of $y \Rightarrow \dfrac{k}{x}$.

In its simplest form, when $k = 1$, the equation becomes $y = \dfrac{1}{x}$. The graph of this function presents some new difficulties which will be apparent on drawing the curve. The following is a table of values formed in the usual way:

x .	$\frac{1}{4}$	$\frac{1}{2}$	1	2	3	4
y .	4	2	1	$\frac{1}{2}$	$\frac{1}{3}$	$\frac{1}{4}$

A similar set of values can be tabulated for negative values of x, the corresponding values of y being negative.

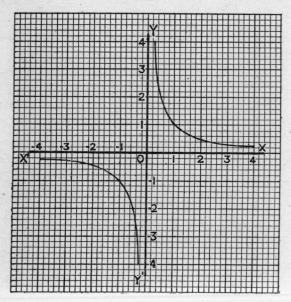

Fig. 48.

The curve, which is called a *hyperbola*, is shown in Fig. 48. It consists of two branches, alike in shape, one for +ve values and the other for —ve values of x. The following important features of this curve should be noted.

(1) As x increases, y or $\frac{1}{x}$ decreases. When x becomes very great, y becomes very small, and the curve approaches very close to the axis of x.

(2) As x decreases, y or $\frac{1}{x}$ increases. When x is very small, y is very large, and the curve approaches very close to the axis of y. Both these features of the curve are repeated for —ve values of x. It may be noted that the curve is symmetrical (i) about the line through the origin making 45° with the x axis—*i.e.*, the line $y = x$; (ii) about a line through the origin at right angles to this—*i.e.*, the line $y = -x$ (see §§ 72 and 73).

For other values of k the curve of $y = \frac{k}{x}$ is always a hyperbola.

174. Other forms of inverse variation.

(1) One quantity may vary inversely as the square of another quantity—*i.e.*, $y \propto \frac{1}{x^2}$, whence $y = \frac{k}{x^2}$. In electricity, for example, the force between two magnetic poles varies inversely as the square of the distance between them. Many other physical laws involve this form of variation.

(2) Another form of variation is that in which one quantity varies inversely as the cube of another—*i.e.*, $y \propto \frac{1}{x^3}$ and $y = \frac{k}{x^3}$.

Generalising, y may vary directly or inversely as any power of x—*i.e.*, $y \propto x^n$ or $y \propto \frac{1}{x^n}$.

In all cases of direct variation the same method, the introduction of the constant k, is followed and the evaluation of k proceeds on the same lines.

175. Worked examples.

Example I. *If y varies as the cube root of x, and if y = 3*

when $x = 64$, *find the formula connecting the variables.*
Hence find x *when* $y = \frac{15}{4}$.

(1) Since
$$y \propto x^{\frac{1}{3}}$$
$$y = kx^{\frac{1}{3}}.$$
$\therefore$
$$k = y \div x^{\frac{1}{3}}.$$

Substituting the given values
$$y = 3 \div 64^{\frac{1}{3}}$$
$$= 3 \div 4 \text{ or } \tfrac{3}{4}.$$
$\therefore$ the law is
$$y = \tfrac{3}{4}\sqrt[3]{x}.$$

(2) When $y = \frac{15}{4}$, we have
$$\tfrac{15}{4} = \tfrac{3}{4}\sqrt[3]{x}.$$
$\therefore$
$$\sqrt[3]{x} = \tfrac{15}{4} \div \tfrac{3}{4}.$$
$$= 5.$$
$\therefore$
$$x = 5^3 = 125.$$

Example 2. *The time of vibration of a simple pendulum*
varies as the square root of its length. If the length of a
pendulum which beats seconds is 1 m, what will be the time of
vibration if its length is increased by 6 cm?

Let l = length of the pendulum in metres and
let t = the time of variation.

Then $\qquad t \propto \sqrt{l}$ and $t = k\sqrt{l}$.
When $l = 1 \cdot 00$, $t = 1 \cdot 00$ and $1 = k\sqrt{1}$, $\therefore k = 1$.
When $l = 1 \cdot 06$, $t = k\sqrt{l} = 1\sqrt{1 \cdot 06} = 1 \cdot 03$.
$\therefore \qquad\qquad t = 1 \cdot 03 \text{ s}.$

Exercise 50.

1. If y is proportional to x^2 and when $x = 15$, $y = 200$,
find the equation connecting x and y. Find y when $x = 8 \cdot 5$.

2. If $y \propto x^3$ and $x = 2$ when $y = 2$, find the law connecting x and y. Then find y when $x = 3$.

3. If $y \propto \dfrac{1}{x^2}$, fill up the blanks in the following table:

x .	1	2	3	
y .			$\frac{4}{3}$	$\frac{1}{3}$

4. If $y \propto \sqrt{x}$ and if $y = 3 \cdot 5$ when $x = 4$, express y in terms of x. What is y when $x = 25$?

5. If $y \propto x^3$ and if $y = 6$ when $x = 4$, find the value of y when $x = 16$. Also find x when $y = 3$.

6. If y is inversely proportional to x, and $x = 5$ when $y = 6$, find the law connecting x and y and find x when $y = 20$.

7. If $y \propto \dfrac{1}{x}$, fill up the gaps in the following table:

x .		$1 \cdot 2$	8	
y .	6		$1 \cdot 5$	$0 \cdot 8$

8. The force F which acts between two magnetic poles is inversely proportional to the square of the distance (d) between them. Express this as a formula if $F = 120$ when $d = 4$.

9. The distance through which a heavy body falls from rest varies as the square of the time taken. A lead ball falls through 490 m in 10 s. How long will it take to fall through $78 \cdot 4$ m?

10. The intensity of illumination given by a projector varies inversely as the square of the distance of its lamp from the screen. If the projector is 20 m from the screen, where must it be placed so that the illumination is 4 times as great?

11. If the distances of an object and of its image, formed by a mirror, are measured from a certain point, it is found that the sum of the distance varies as their product. If the image distance is 120 mm when the object distance is 300 mm, calculate the image distance when the object distance is 540 mm.

12. The square of the speed of a particle varies as the cube of its distance from a fixed point. If this distance is increased by 1·2 per cent., what is the approximate percentage increase in the speed?

13. A clock keeps accurate time at 10° C, but gains as the temperature falls and vice-versa, the rate of gain or loss varying as the square of the number of °C between the actual temperature and 10° C. If it gains 2 s per day when the temperature is 5° C, how much does it lose (to the nearest s) in 4 days when the temperature is 42° C?

14. If $y \propto x^{1·4}$ and if $y = 354·5$ when $x = 15$, find the law connecting y and x.

176. Functions of more than one variable.

Example 1.

It is proved in elementary Geometry that the area of a triangle is given by the formula

$$A = \tfrac{1}{2}bh$$

where A = the area of the triangle
  b = the length of a base
  h = the corresponding altitude.

Both b and h are variables, and the value of A depends on them both.

∴ A is a function of the two variables b and h.

In any triangle let the height remain constant, but the base variable, then if the *base* be doubled, the area will be doubled. If the base remain constant and the *height* be trebled, the area will be trebled.

Now suppose both base and height to vary; let the base be doubled and the height trebled, then the area will be 2×3—*i.e.*, 6 times greater.

We can infer, then, that if both base and height vary, the area varies as the product of base and height—*i.e.*,

$$A \propto b \times h.$$
∴ $$A = kbh.$$

From geometrical considerations we know that in this case $k = \tfrac{1}{2}$.

Example 2. In § 172 it was stated that the volume of a given mass of gas, at a constant temperature, varies inversely as the pressure on it. It can also be shown by experiment that if the pressure be kept constant and the temperature varied, then the volume varies directly as the absolute temperature. If both temperature and pressure vary, then the volume varies **directly as the absolute temperature and inversely as the pressure.**

Let v = volume

   ,, T = absolute temperature

   ,, p = pressure

then
$$v \propto \frac{T}{p}.$$

$\therefore$
$$v = k \times \frac{T}{p}.$$

Example 3. If an electric current passes through a wire, it encounters resistance. This resistance **varies as the length of the wire** (§ 162), and in a wire of given length— *i.e.*, constant—it **varies inversely** as the cross section of the wire (§ 172).

$\therefore$ if R = the resistance

    l = the length

    A = the cross section

then R varies directly as l and inversely as A.

Thus
$$R \propto \frac{l}{A}$$

and
$$R = k \times \frac{l}{A}.$$

177. Joint variation.

The variation of a quantity due to two or more variables is sometimes called **joint variation,** and the quantity is said to **vary jointly as their product.**

In dealing with problems involving joint variation, the same procedure with regard to the constant of variation and its determination is followed as in the previous case. The following examples will serve to illustrate it.

178. Worked example.

Example 1. *A quantity represented by y varies directly as x and inversely as z^3. It is known that when $x = 15$, $z = 12$ and $y = \frac{1}{36}$. Find the law connecting the quantities.*

We are given that $\quad y \propto \dfrac{x}{z^3}$.

$\therefore \qquad\qquad\qquad y = k \times \dfrac{x}{z^3}$.

Substituting the given values

$$\tfrac{1}{36} = k \times \frac{15}{12^3}.$$

$\therefore \qquad\qquad k = \tfrac{1}{36} \div \frac{15}{12^3} = \tfrac{16}{5}.$

$\therefore$ the law is $\qquad y = \tfrac{16}{5} \dfrac{x}{z^3}$.

Example 2. *The force between two magnetic poles varies jointly as the product of their strength and inversely as the square of the distance between them. If two poles of strength 8 and 6 units repel one another with a force of 3 N when placed 4 m apart, with what force will two poles whose strengths are 5 and 9 units repel one another when 2 m apart?*

Let $\quad F$ = the force
„ $\quad m_1, m_2$ = the pole strengths
„ $\qquad d$ = the distance apart

then $\qquad\qquad F \propto \dfrac{m_1 m_2}{d^2}$.

$\therefore \qquad\qquad F = k \times \dfrac{m_1 m_2}{d^2}$.

Substituting the given values

$$3 = k \times \frac{8 \times 6}{4^2}.$$

$\therefore \qquad\qquad k = \frac{3 \times 4^2}{8 \times 6} = 1.$

$\therefore \qquad\qquad F = \dfrac{m_1 m_2}{d^2}$.

In the second case $\quad F = \dfrac{5 \times 9}{4} = 11\cdot25 \text{ N}$.

Exercise 51.

1. Express the following statements in the form of equations:

(a) y varies jointly as x and z.

(b) y varies directly as x and inversely as the square of z.

(c) y varies directly as the square root of x and inversely as z.

(d) The volume (V) of a cylinder varies jointly as the height (h) and the square of the radius of the base (r).

(e) The weight (W) which can be carried safely by a beam varies inversely as the length (l), directly as the breadth (b) and directly as the square of the depth (d).

(f) y varies directly as the square of x and inversely as the cube root of z.

2. If y varies directly as x and inversely as z, and if $y = 10$ when $x = 8$ and $z = 5$, find the law connecting x, y and z. Also find y when $x = 6$ and $z = 2\cdot5$.

3. If y varies jointly as x and z^2, and if $y = 13\frac{1}{3}$ when $x = 2\cdot5$ and $z = \frac{4}{3}$, find the law connecting the variables. Also find x when $z = \frac{3}{2}$ and $y = 54$.

4. y varies directly as x^2 and inversely as $\sqrt{z}$. When $x = 8$ and $z = 25$, $y = 16$. Find y when $x = 5$ and $z = 9$.

5. The load that a beam of given depth can carry is directly proportional to the breadth and inversely proportional to the length. If a beam of length 7 m and width 175 mm can support a load of 4 t, what load can be supported by a beam of the same material 5 m long and 250 mm wide?

6. If z varies as x^2 and inversely as y^2, and if $z = 4$ when $x = 8$ and $y = -0\cdot5$, find z when $x = -2$ and $y = 0\cdot25$.

7. The number of heat units (H) generated by an electric current varies directly as the time (t) and the square of the voltage (E) and inversely as the resistance (R). If $H = 60$, when $t = 1$, $E = 100$ and $R = 40$ find the law connecting them.

Also find:

(1) The value of H when $E = 200$, $R = 120$ and $t = 300$.

(2) The value of t when $E = 120$, $R = 90$ and $H = 5760$.

8. The pressure in a test apparatus varies directly as the absolute temperature and inversely as the volume of gas. If the temperature increases by 2 per cent. and the volume decreases by 2 per cent., what is the change in the pressure?

CHAPTER XIX

THE DETERMINATION OF LAWS

179. Laws which are not linear.

IN the preceding chapter we considered the determinations of laws which were linear, and which were arrived at by using experimental data.

But such laws are not always represented by straight lines. They may involve powers of a variable such as were considered in § 169 and onwards. In these cases, when the results of the experiments are plotted, they will lie on a portion of a curve which might be one of those illustrated in the preceding chapter or of many others. In practice, when only a small portion of the curve can be drawn, it is impossible to identify what curve it is.

There are two devices, however, by means of which a straight line can be obtained instead of a curve. The identification can then be made by the methods previously considered.

180. $y = ax^n + b$. Plotting against a power of a number.

If the law which we require to find is of the form $y = ax^n + b$ or $y = ax^n$, where n is known, then we plot corresponding values of y and x^n instead of y and x. The resulting graph will be a straight line.

Let us consider as an example $y = 2x^2$.

The graph of this function was found in § 111 to be a parabola. The values given to x were 0, 1, 2, 3 . . .

If we plot corresponding values of y and x^2 the table of values will be

x^2 .	0	1	2	3	4
y .	0	2	4	6	8

and so on. The values of x are not shown in the table.

The resulting graph is a straight line passing through the origin as shown in Fig. 49. It is the same as $y = 2x$ when y is plotted against x.

If the equation is $y = 2x^2 + 3$ on plotting in the same way, that is y against x^2, the resulting line will not pass through the origin but will have an intercept of 3 units on the y axis. It will be the same as $y = 2x + 3$, when y is plotted against x.

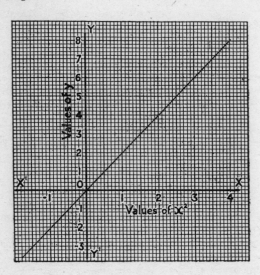

Fig. 49.

The same procedure will be followed with any other power of x. Thus in $y = 2x^3 + 5$, y is plotted against x^3.

In general for the function $y = ax^n + b$ plot y against x^n.

If $x^n = z$ then the equation takes the form of

$$y = az + b.$$

This is the equation of a straight line. The graph, while not showing the relation between y and x as graphs usually do, will make it possible to find the values of a and b by methods previously given.

Worked example.

Two variables, x and y, are thought to be connected by a law of the form of $y = ax^2 + b$. The following values of x and y are known. Find the law connecting the variable.

x .	0·5	1	1·5	2	2·5
y .	−9·25	−7	−3·25	2	8·75

y must be plotted against x^2; we therefore calculate the following tables of corresponding values of x^2 and y.

x^2 .	0·25	1	2·25	4	6·25
y .	−9·25	−7	−3·25	2	8·75

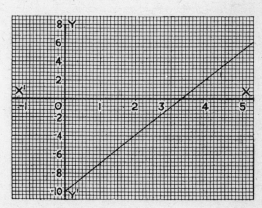

Fig. 50.

Plotting y against x^2, the resulting graph is as shown in Fig. 50.

This is a straight line, and the values of a and b can be found by the method of § 168.

By inspection of the graph, the intercept on the y axis (*i.e.*, b) is −10, and a, the gradient of the line, is 3.

∴ the law is $y = 3x^2 - 10.$

181. $y = ax^n$. Use of logarithms.

As was pointed out, the previous method can be used only when the power of x involved is known. If, however, it is not known, and the law is of the form $y = ax^n$, this can be reduced to the form of the equation of a straight line by taking logarithms.

If $\qquad\qquad y = ax^n$

then $\qquad\qquad \log y = n \log x + \log a.$

Comparing this with the standard form of the equation of a straight line, viz., $y = ax + b$, it is seen to be linear and of the same form, $\log y$ taking the place of y, and $\log x$ taking the place of x.

The constants to be determined are now n and $\log a$. Therefore we plot the graph of

$$\log y = n \log x + \log a.$$

From this graph n and $\log a$ can be found in the same way as a and b in the standard form. When $\log a$ is known we find a from the tables and the law can be written down.

It is possible to deal with only the simpler cases in this book.

182. Worked example.

Two variables, x and y, are connected by a law of the form $y = ax^n$. The following table gives corresponding values of x and y. Find the law connecting these.

x .	18	20	22	24	25
y .	623	863	1160	1519	1724

Since the law connecting these is of the form

$$y = ax^n$$

taking logs $\qquad\qquad \log y = n \log x + \log a.$

Tabulating values of $\log x$ and $\log y$, we get the following:

$\log x$.	1·255	1·301	1·342	1·380	1·398
$\log y$.	2·795	2·936	3·065	3·182	3·236

The logs correspond in order to the numbers in the column above, and are calculated approximately to 3 places of decimals.   The graph is the straight line shown in Fig. 51.

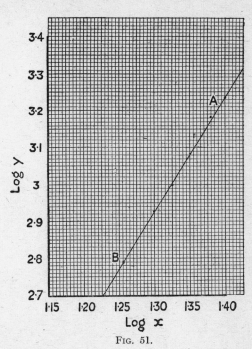

Fig. 51.

Selecting the points A and B on the straight line, we substitute their co-ordinates in turn in the equation

$$\log y = n \log x + \log a.$$

Thus we get the equations

$$3 \cdot 236 = 1 \cdot 398n + \log a.$$
$$2 \cdot 795 = 1 \cdot 255n + \log a.$$

Subtracting, $0 \cdot 441 = 0 \cdot 143n.$

$\therefore \qquad\qquad n = \dfrac{0 \cdot 441}{0 \cdot 143}$

$\qquad\qquad\quad = 3 \cdot 1$ approx.

Substituting for n in equation above

$$2 \cdot 795 = (3 \cdot 1 \times 1 \cdot 255) + \log a.$$
$$\therefore \qquad \log a = 2 \cdot 795 - 3 \cdot 891$$
$$= \bar{2} \cdot 904.$$
$$\therefore \qquad a = 0 \cdot 08 \text{ approx.}$$

$\therefore$ the law connecting y and x is

$$y = 0 \cdot 08 x^{3 \cdot 1}.$$

Exercise 52.

1. The variables x and y are connected by a law of the form $y = ax^2 + b$. The following corresponding values of x and y are known. Find the law.

x .	0·5	1	1·5	2	2·5
y .	4·5	9	16·5	27	40·5

2. The following table gives related values of x and y. Determine whether these values are connected by an equation of the form $y = ax^2 + b$ and, if so, find the values of a and b.

x .	4	5	6	7	8	9
y .	14·3	18	22·5	28	34·5	41·5

3. The following values of R and V are possibly connected by a law of the type $R = aV^2 + b$. Test if this is so and find the law.

V .	12	16	20	22	24	26
R .	6·44	7·56	9	9·84	10·76	11·76

4. H is connected with V by an equation of the form $H = aV^3 + b$. The following corresponding values are known. Find the values of a and b.

V .	10	12	14	15	17
H .	1500	2300	3400	4000	5700

5. The following corresponding values of x and y were measured. There may be errors of observation. Test if there is a probable law $y = a + bx^2$ and, if this is the case, find the probable values of a and b.

x .	1	1·5	2	2·30	2·50	2·70	2·80
y .	0·77	1·05	1·50	1·77	2·03	2·25	2·42

6. In measuring the resistance, R ohms, of a carbon-filament lamp at various voltages, V, the following results were obtained:

V .	60	70	80	90	100	120
R .	70	67·2	65	63·3	62	60

The law is thought to be of the form $R = \dfrac{a}{V} + b$. Test this and find a and b.

7. The values of x and y in the following table are connected by a law of the form $y = ax^n$. Find a and n and so determine the law.

x .	2	3	4	5
y .	2	6·75	16	31·25

8. The following table gives corresponding values of two

variables x and y. The law which connects them is of the form $y = ax^n$. Find this law.

x .	2	3	5	6	8	10
y .	4·24	5·20	6·71	7·35	8·49	9·49

9. The following values of H and Q are connected by a law of the type $Q = aH^n$. Find a and n.

H .	1·2	1·6	2·0	2·2	2·5	3
Q .	6·087	6·751	7·316	7·571	7·927	8·467

10. Two quantities x and y are connected by an equation of the form $y = ax^n$. The following table gives corresponding values of the variables. Determine a and n.

x .	3	3·5	4	4·5	5
y .	6·19	6·79	7·35	7·89	8·40

CHAPTER XX

RATIONAL AND IRRATIONAL NUMBERS. SURDS

183. Rational and irrational numbers.

A NUMBER which is either an integer or can be expressed as the **ratio** of two integers is called a **rational number**.

A number which cannot be expressed as an integer or as a ratio with a finite number of figures is called an **irrational number**.

Thus $\sqrt{2}$ cannot be expressed as a fraction or a ratio with a finite number of figures. It can, however, be expressed as a decimal to any required degree of accuracy.

Thus to 4 significant figures $\sqrt{2} = 1\cdot414$.

,,   6   ,,   ,, $\sqrt{2} = 1\cdot414\ 21$ and so on.

But it can be proved that there is no limit to the number of figures in the decimal place, that is to say the number of figures is **not finite**. It is therefore an irrational number.

Other roots such as $\sqrt{3}$, $\sqrt{11}$, $\sqrt{19}$, $\sqrt[3]{4}$, . . . are examples of irrational numbers. A root of a rational number which is irrational as these are is called a **surd**.

There are also other numbers, which do not involve roots which are irrational. Thus the ratio of the circumference of a circle to its diameter, which we denote by the symbol π, cannot be expressed by a ratio with a finite number of figures. It is often expressed roughly by $\frac{22}{7}$, or $3\cdot1416$ to 5 significant figures, but it has been proved that there is no limit to the number of figures in the decimal part. Computers have evaluated π to an immense number of places, but they are, of course, no nearer an end.

Such numbers as this are irrational, but not surds. They are also called **incommensurable**.

184. Irrational numbers and the number scale.

Since such numbers cannot be expressed with absolute accuracy it is not possible to assign an exact position for them in the complete number scale (§ 36.) We can, however, state to any required degree of accuracy the limits between which they lie. Thus $\sqrt{2}$ lies on the scale between 1·414 and 1·415

or more accurately between 1·4142 and 1·4143
or more accurately still between 1·414 21 and 1·414 22
and so on.

185. Geometrical representation of surds.

It may be noted, however, that it is theoretically possible in many cases to obtain by geometrical constructions, straight lines which do represent surds accurately.

For example, we know from Geometry that the square on the hypotenuse of a right-angled triangle is equal to the sum of the squares on the sides containing the right angle. Consequently if a right-angled triangle be constructed, the sides of which are of unit length, the length of the hypotenuse must be $\sqrt{1^2 + 1^2}$—*i.e.*, $\sqrt{2}$.

Similarly the hypotenuse of a right-angled triangle of sides 1, and $\sqrt{2}$ units will be $\sqrt{3}$ units. In this way it is possible to represent many surds by straight lines. The lengths thus obtained can be marked on the number scale, but in practice no high degree of accuracy can be obtained in the construction of them.

186. Operations with surds.

It is the custom in Algebra to classify such a number as $\sqrt{a}$ as a surd, though until a numerical value has been assigned to a we cannot say whether or no it is irrational. For purposes of operation, however, it is treated as a surd.

In operating with surds one principle is fundamental. **Surds must obey the laws of Algebra** as formulated for rational numbers. Since surds can also be written as powers with fractional indices, *e.g.* $\sqrt{2} = 2^{\frac{1}{2}}$, we can operate with surds as with these powers, according to the laws of indices.

For example, just as $(a + b)^{\frac{1}{2}}$ is not equal to $a^{\frac{1}{2}} + b^{\frac{1}{2}}$, so $\sqrt{a + b}$ is *not* equal to $\sqrt{a} + \sqrt{b}$. In this respect the root sign has the same effect as a bracket; the expression under it must be regarded as a whole.

(1) Multiplication.

$$\sqrt{a} \times \sqrt{b} = a^{\frac{1}{2}} \times b^{\frac{1}{2}} = (ab)^{\frac{1}{2}} = \sqrt{a \times b}.$$

Thus:

$$\sqrt{3} \times \sqrt{7} = \sqrt{3 \times 7} = \sqrt{21}$$
$$\sqrt{2}(\sqrt{5} - \sqrt{3}) = (\sqrt{2} \times \sqrt{5}) - (\sqrt{2} \times \sqrt{3}) = \sqrt{10} - \sqrt{6}$$
$$4\sqrt{7} = \sqrt{16} \times \sqrt{7} = \sqrt{16 \times 7} = \sqrt{112}$$
$$3x^2\sqrt{y} = \sqrt{9x^4} \times \sqrt{y} = \sqrt{9x^4y}$$
$$(\sqrt{a} + \sqrt{b})^2 = (\sqrt{a})^2 + 2(\sqrt{a} \times \sqrt{b}) + (\sqrt{b})^2$$
$$= a + b + 2\sqrt{ab}$$
$$(\sqrt{a} - \sqrt{b})^2 = a + b - 2\sqrt{ab}$$
$$(\sqrt{a} + \sqrt{b})(\sqrt{a} - \sqrt{b}) = (\sqrt{a})^2 - (\sqrt{b})^2 = a - b$$
$$(\sqrt{5} + 7)(\sqrt{3} - 2) = (\sqrt{5} \times \sqrt{3}) - 2\sqrt{5} + 7\sqrt{3} - 14.$$

By using the above and the converse rules we obtain useful transformations in operations.

Thus:

$$\sqrt{1000} = \sqrt{100 \times 10} = \sqrt{100} \times \sqrt{10} = 10\sqrt{10}$$
$$\sqrt{72} = \sqrt{36 \times 2} = \sqrt{36} \times \sqrt{2} = 6\sqrt{2}$$
$$\sqrt{9a^3b^2} = \sqrt{9 \times a^2 \times a \times b^2} = 3ab\sqrt{a}.$$

The above transformations may also be employed to simplify expressions involving surds.

Examples

(1) $\sqrt{5} + \sqrt{20} = \sqrt{5} + \sqrt{4 \times 5} = \sqrt{5} + 2\sqrt{5} = 3\sqrt{5}.$

(2) $\sqrt{27} - \sqrt{75} + \sqrt{48}$
$$= \sqrt{9 \times 3} - \sqrt{25 \times 3} + \sqrt{16 \times 3}$$
$$= 3\sqrt{3} - 5\sqrt{3} + 4\sqrt{3}$$
$$= 2\sqrt{3}.$$

(2) **Rationalisation.** The evaluation of a number such as $\dfrac{1}{\sqrt{2}}$ will be easier, and likely to be more accurate, if the fraction can be transformed so that we multiply by the surd and do not divide by it. This can be done by the following transformation:

$$\frac{1}{\sqrt{2}} = \frac{\sqrt{2}}{\sqrt{2} \times \sqrt{2}} = \frac{\sqrt{2}}{2}.$$

Similarly

$$\frac{\sqrt{5}}{2\sqrt{3}} = \frac{\sqrt{5} \times \sqrt{3}}{2\sqrt{3} \times \sqrt{3}} = \frac{\sqrt{15}}{2 \times 3} = \frac{\sqrt{15}}{6}.$$

By this transformation the denominator is changed from an irrational number to a rational one.

This is called **rationalising the denominator**.

If the denominator is a binomial expression the method is slightly more difficult. The procedure is indicated in the following examples:

Example 1. *Rationalise the denominator of* $\dfrac{1}{\sqrt{5} - \sqrt{2}}$.

Since $(a - b) \times (a + b) = a^2 - b^2$ and $(\sqrt{a} + \sqrt{b}) \times (\sqrt{a} - \sqrt{b}) = a - b$ (see above), then, if the denominator is multiplied by $\sqrt{5} + \sqrt{2}$, the surds will disappear from it.

Thus

$$\frac{1}{\sqrt{5} - \sqrt{2}} = \frac{\sqrt{5} + \sqrt{2}}{(\sqrt{5} - \sqrt{2})(\sqrt{5} + \sqrt{2})}$$
$$= \frac{\sqrt{5} + \sqrt{2}}{5 - 2} = \frac{\sqrt{5} + \sqrt{2}}{3}.$$

Example 2. *Simplify* $\dfrac{\sqrt{5} - 1}{\sqrt{5} + 1}$.

To rationalise the denominator it must be multiplied by $\sqrt{5} - 1$.

$$\therefore \quad \frac{\sqrt{5} - 1}{\sqrt{5} + 1} = \frac{(\sqrt{5} - 1)(\sqrt{5} - 1)}{(\sqrt{5} + 1)(\sqrt{5} - 1)} = \frac{(\sqrt{5} - 1)^2}{(\sqrt{5})^2 - (1)^2}$$
$$= \frac{5 - 2\sqrt{5} + 1}{5 - 1}$$
$$= \frac{6 - 2\sqrt{5}}{4} = \frac{3 - \sqrt{5}}{2}.$$

Exercise 53.

1. Express the following as complete square roots, thus: $3\sqrt{7} = \sqrt{9 \times 7} = \sqrt{63}.$

(1) $5\sqrt{6}$.   (2) $12\sqrt{2}$.

(3) $10\sqrt{5}$.   (4) $4\sqrt{13}$.

(5) $2a\sqrt{b}$.   (6) $3x^2\sqrt{y}$.

2. Express the following with the smallest possible number under the root sign, in each case:

(1) $\sqrt{800}$.   (2) $\sqrt{320}$.

(3) $\sqrt{108}$.   (4) $\sqrt{5000}$.

(5) $\sqrt{375}$.   (6) $\sqrt{7200}$.

(7) $\sqrt{24a^3b^2}$.   (8) $\sqrt{18x^4y^3z}$.

(9) $\sqrt{75a^3b^3c^3}$.   (10) $\sqrt{1000p^2q}$.

(11) $\sqrt{a^2b(a-b)^3}$.   (12) $\sqrt{25xy^3(x-2y)^2}$.

3. Simplify:

(1) $\sqrt{3} \times \sqrt{15}$.   (2) $\sqrt{14} \times \sqrt{7}$.

(3) $\sqrt{32} \times \sqrt{24}$.   (4) $2\sqrt{5} \times 5\sqrt{2}$.

(5) $\sqrt{2}(2\sqrt{2}-1)$.   (6) $\sqrt{7}(\sqrt{14}+\sqrt{2})$.

(7) $\sqrt{a+b} \times \sqrt{a^2-b^2}$.

(8) $\sqrt{3(x-2y)} \times \sqrt{6(x^2-4y^2)}$.

4. Multiply the following:

(1) $(\sqrt{2}-1)(\sqrt{2}+3)$

(2) $(2\sqrt{3}+\sqrt{2})(\sqrt{3}-2\sqrt{2})$.

(3) $(\sqrt{2}-1)^2$.   (4) $(2\sqrt{5}+\sqrt{3})^2$.

(5) $(\sqrt{7}-5\sqrt{2})^2$.

(6) $(\sqrt{5}+\sqrt{3})(\sqrt{5}-\sqrt{3})$

(7) $(2\sqrt{10}-\sqrt{2})(2\sqrt{10}+\sqrt{2})$.

(8) $(1+10\sqrt{3})^2$.   (9) $(\sqrt{a}-5)(\sqrt{a}+5)$.

(10) $(\sqrt{27}+\sqrt{6})(\sqrt{27}-\sqrt{6})$.

5. Simplify the following by rationalising the denominator:

(1) $\dfrac{2}{\sqrt{3}}$.   (2) $\dfrac{1}{\sqrt{5}}$.

(3) $\dfrac{3}{2\sqrt{7}}$.   (4) $\dfrac{1}{3\sqrt{10}}$.

(5) $\dfrac{1}{\sqrt{32}}$.   (6) $\dfrac{12}{\sqrt{20}}$.

(7) $\dfrac{\sqrt{2}}{\sqrt{3}}$.   (8) $\dfrac{2\sqrt{5}}{\sqrt{7}}$.

(9) $\dfrac{1}{\sqrt{5}-1}$.   (10) $\dfrac{3}{\sqrt{7}+\sqrt{2}}$.

(11) $\dfrac{3}{2\sqrt{3} - \sqrt{2}}.$

(12) $\dfrac{5}{2\sqrt{7} - 3\sqrt{3}}.$

(13) $\dfrac{\sqrt{3} + \sqrt{2}}{\sqrt{3} - \sqrt{2}}.$

(14) $\dfrac{\sqrt{7} - \sqrt{5}}{\sqrt{7} + \sqrt{5}}.$

(15) $\dfrac{\sqrt{7}}{3\sqrt{7} - 1}.$

(16) $\dfrac{1}{\sqrt{2} - 1} + \dfrac{1}{\sqrt{2} + 1}.$

(17) $\dfrac{1}{\sqrt{5} + \sqrt{3}} - \dfrac{1}{\sqrt{5} - \sqrt{3}}.$

(18) $\dfrac{4}{6 - \sqrt{5}}.$

CHAPTER XXI

SERIES

Arithmetical and Geometrical Progressions

187. Meaning of a series.

A series is a succession of numbers each of which is formed according to a definite law, which is the same throughout the series.

The ordinary numbers $1, 2, 3, 4, \ldots$ constitute a series, each term of which is greater by unity than the one which immediately precedes it.

$5, 9, 13, 17, \ldots$ is a series, each term of which is greater by 4 than the one which immediately precedes it.

$2, 4, 8, 16, \ldots$ is a series in which each term is twice the one which immediately precedes it.

$1, \frac{1}{2}, \frac{1}{3}, \frac{1}{4}, \ldots$ is a series in which the terms are the reciprocals of $1, 2, 3, 4, \ldots$

$1^2, 2^2, 3^2, \ldots$

$1^3, 2^3, 3^3, \ldots$

are series, the construction of each of which is obvious.

188. The formation of a series.

Series are of great importance in modern Mathematics, but in this book it is possible to deal with only a few simple cases.

The two most important things to be known about a series are:

(1) **The law of its formation.** If this is known, it is possible to find any term in the series.

(2) **The sum of a given number of terms of the series.**

In this connection it is necessary to consider what is the nature of the sum when the number of terms is very great.

If the series is one in which the terms increase numerically such as

$$2, 5, 8, 11, \ldots$$

or

$$1, 3, 9, 27 \ldots$$

it is clear that the more terms which are taken, the greater will be the sum. But if the series is one in which the terms decrease as the number of terms increases, such as $1, \frac{1}{3}, \frac{1}{9}, \frac{1}{27}, \ldots$, it is not always easy to discover what the sum will be when the number of terms is very great. This is a matter which will be considered later.

189. Arithmetic series.

An Arithmetic series, or Arithmetic progression, is one in which each term is formed from that immediately preceding it by adding or subtracting a constant number.

The number thus added or subtracted is called the **common difference** of the series.

Examples.

(1) $7, 13, 19, 25, \ldots$ (common difference 6).
(2) $6, 4, 2, 0, -2, \ldots$ (common difference -2).

In general if three numbers a, b, c, are in Arithmetic progression (denoted by A.P.) then

$$b - a = c - b.$$

190. Any term in an Arithmetic series.

Let $a =$ the first term of a series

,, $d =$ the common difference (+ve or −ve)

then the series can be written

$$a, a + d, a + 2d, a + 3d, \ldots$$

It is evident that the *multiple of d* which is added to a to produce any term is *one less than the number of the term in the sequence*.

Thus the fourth term is $a + (4 - 1)d = a + 3d$. Hence if the number of any term be denoted by n

then $\qquad$ **nth term $= a + (n - 1)d$.**

Examples.

(1) In the series $7, 10, 13, \ldots$ the common difference is 3.

$\therefore$ the tenth term is $7 + (10 - 1)3 = 34$.

$\therefore$,, nth ,, $7 + (n - 1)3$.

(2) In the series 6, 2, $- 2$, $- 6$, . . . $d = - 4$.

∴ the nth term $= 6 + (n - 1)(- 4) = 6 - (n - 1)4$.

∴ the eighth term $= 6 + 7(- 4) = 6 - 28 = - 22$.

191. The sum of any number of terms of an Arithmetic series.

Using the following symbols in addition to those used previously:

Let n = number of terms whose sum is required

,, s = the sum of n terms

,, l = the last term

then by the previous formula

$$l = a + (n - 1)d.$$

Now

$$s = a + (a + d) + (a + 2d) + . . . + (l - d) + l.$$

Reversing the series

$$s = l + (l - d) + (l - 2d) + . . . + (a + d) + a.$$

Adding the corresponding terms of the two sets, each pair gives $(a + l)$.

$$2s = (a + l) + (a + l) + (a + l) + . . . + (a + l) + (a + l)$$
$$= (a + l) \times n, \text{ since there are } n \text{ terms and } \therefore n \text{ pairs.}$$

$$\therefore \qquad s = \frac{n(a + l)}{2}.$$

Since $l = a + (n - 1)d$, on substituting for l in the last result

$$s = \frac{n\{a + a + (n - 1)d\}}{2}.$$

$$\therefore \qquad s = \frac{n}{2}\{2a + (n - 1)d\}.$$

This formula, like all other formulae, may be used not only to find s, but also any of the other numbers n, a, or d.

To find a and d offers no difficulty, but if n is required it will be seen that a **quadratic equation** will result. Since there are two roots to every quadratic equation, two values of n will always be found. In some cases only one root is admissible; in others both roots provide solutions.

For example, in a series involving negative terms such as

$$9 + 7 + 5 + 3 + 1 - 1 - 3$$

it is seen that the sum of 7 terms is the same as the sum of 3 terms. In other cases it will be evident that one of the roots is inadmissible.

192. Arithmetic mean.

If three numbers are in Arithmetic progression, the middle one is called the Arithmetic mean of the other two.

Let a, b, c, be three numbers in A.P.
then by the definition of § 189

$$b - a = c - b$$
$$2b = a + c.$$
$$\therefore \qquad b = \frac{a + c}{2}.$$

It will be seen that the Arithmetic mean of two numbers is the same as their average.

It is usual also to speak of inserting Arithmetic means between two numbers, by which is meant that they, together with the two given, form a series of numbers in A.P.

Example. Insert three Arithmetic means between 4 and 20.

If these be a, b, c, then 4, a, b, c, 20 are in A.P., five terms in all.

Using $l = a + (n - 1)d$ for the fifth term 20

$$20 = 4 + (5 - 1)d, \text{ whence } d = 4.$$

$\therefore$ the five terms are 4, 8, 12, 16, 20.

193. Worked examples.

Example 1. *The sum of an A.P. of 8 terms is 90 and the first term is 6. What is the common difference?*

Using $s = \frac{n}{2}\{2a + (n - 1)d\}$ and substituting given values $\quad 90 = \frac{8}{2}\{(2 \times 6) + (8 - 1)d\}$

$$90 = 4(12 + 7d) = 48 + 28d.$$
$$\therefore \qquad 28d = 42$$
$$\text{and} \qquad d = 1{\cdot}5.$$

Example 2. *How many terms of the series* 3, 6, 9, . . .
must be taken so that their sum is 135?

Using $\qquad s = \frac{n}{2}\{2a + (n-1)d\}$

and substituting, $135 = \frac{n}{2}\{6 + (n-1)3\}$.

$\therefore \qquad\qquad 270 = n(3 + 3n)$
or $\qquad\qquad 270 = 3n + 3n^2$.
$\therefore \quad n^2 + n - 90 = 0$.

Factorising $\qquad (n - 9)(n + 10) = 0$
$\therefore \qquad\qquad\qquad\qquad n = 9 \text{ or } -10$.

The root -10 is inadmissible as having no meaning in
this connection.

$\therefore$ the solution is $n = 9$.

Exercise 54.

1. Write down the next three terms of the following
series:

(1) 5, 7·5, 10, . . .
(2) 12, 8, 4, . . .
(3) $(a + 3b)$, $(a + b)$, $(a - b)$.
(4) 2·7, 4, 5·3.
(5) $x - y$, x, $x + y$.

2. Find the fifth and eighth terms of the series whose
first term is 6, and common difference 1·5.

3. Find the $2p$th term of the series whose first term is 6
and common difference 2.

4. Find the nth term of the series whose first term is
$(x + 2)$ and common difference 3.

5. Find the twenty-fifth term of the series 0·6, 0·72,
0·84, . . .

6. The fourth term of an A.P. is 11 and the sixth term
17. Find the tenth term.

7. The fifth term of an A.P. is 11 and the ninth term is 7.
Find the fourteenth term.

8. Which term of the series 2·3, 4·2, 6·1, . . . is 36·5?

9. Find the sums of the following series:

(a) 15, 16·5, 18, . . . to ten terms.
(b) 9, 7, 5, . . . to eight terms.
(c) 0·8, 0·6, 0·4, . . . to nine terms.
(d) $2\frac{2}{3}$, $3\frac{1}{2}$, $4\frac{1}{3}$, . . . to twenty-seven terms.

10. How many terms of the series 10, 12, 14, . . . must be taken so that the sum of the series is 252?

11. How many terms of the series 24, 20, 16, . . . must be taken so that the sum of the series is 80?

12. Find the thirtieth term and the sum of thirty terms of the series 4, 8, 12, . . .

13. A contractor agrees to sink a well 250 m deep at a cost of £2·70 for the first metre, £2·85 for the second metre, and an extra 15p for each additional metre. Find the cost of the last metre and the total cost.

14. A parent places in the savings bank 25p on his son's first birthday, 50p on his second, 75p on his third, and so on, increasing the amount by 25p on each birthday. How much will be saved up when the boy reaches his sixteenth birthday, the latter inclusive?

---

194. Harmonic progression.

A series of numbers is said to be in harmonic progression (H.P.) if their reciprocals form a series in arithmetic progression.

Thus the series 1, 3, 5, 7, . . . are in A.P.

∴   ,,   ,,   1, $\frac{1}{3}$, $\frac{1}{5}$, $\frac{1}{7}$, . . .   ,,   H.P.

This series is important in the theory of sound.

It is not possible to obtain a formula for the sum of n terms of an H.P., but many problems relating to such a series can be solved by using the corresponding arithmetic series.

Harmonic mean.—The harmonic mean of two numbers may be found as follows:

Let a and b be the numbers.
   ,, H be their harmonic mean,
i.e., a, H, b are in H.P.

Then $\dfrac{1}{a}$, $\dfrac{1}{H}$, $\dfrac{1}{b}$ are in A.P.

$$\therefore \qquad \frac{1}{H} = \frac{1}{2}\left(\frac{1}{a} + \frac{1}{b}\right) \quad (\S 192)$$

$$= \frac{a+b}{2ab}.$$

$$\therefore \qquad H = \frac{2ab}{a+b}.$$

195. Geometric series, or geometric progression.

A geometric series is one in which the ratio of any term to that which immediately precedes it is constant for the whole series.

This ratio is called the **common ratio** of the series. It may be positive or negative. Thus each term of the series can be obtained by multiplying the term which precedes it by the common ratio.

Examples.

(1) 1, 2, 4, 8, . . .   (common ratio 2).
(2) 1, $\frac{1}{2}$, $\frac{1}{4}$, $\frac{1}{8}$, . . .   (,, $\frac{1}{2}$).
(3) 2, $-$ 6, 18, $-$ 54, . . . (,,   ,, $-$ 3).
(4) R, R^2, R^3, R^4, . . . (,,   ,, R).

If three numbers a, b, c are in geometric progression (G.P.) then $\dfrac{b}{a} = \dfrac{c}{b}$.

This is the test to apply in order to find if numbers are in G.P.

General form of a geometric series.

Let a = 1st term.
   ,, r = common ratio.
Then the series is a, ar, ar^2, ar^3, . . .

196. Connection between a geometric series and an arithmetic series.

In the geometric series

$$a,\ ar,\ ar^2,\ ar^3,\ \ldots$$

take logs of each term. We then get the series:

$log\ a$, $log\ a + log\ r$, $log\ a + 2\ log\ r$, $log\ a + 3\ log\ r$, . . .

This is an arithmetic series in which the first term is log a, and the common difference is log r.

∴ *the logarithms of the terms of a G.P. form a series in A.P.*

197. General term of a geometric series.

Examining the series a, ar, ar^2, ar^3, . . . it will be seen that each term of the series is the product of a and **a power of r the index of which is one less than the number of the term.**

∴ if n = any term

then the nth term = ar^{n-1}.

If r is negative, $n - 1$ being alternately odd and even, the terms will be alternately negative and positive, assuming a to be positive.

When $n - 1$ is even, n is odd, and the nth-term is +ve.

,, $n - 1$ is odd, n is even,   ,,   ,,   is −ve.

Worked examples.

Example 1. *Find the seventh term of the series*

$$3, 6, 12, . . .$$

In this series $r = 2$, so using the formula

$$n\text{th term} = ar^{n-1}$$
$$\text{the seventh term} = ar^6 = 3 \times 2^6 = 3 \times 64$$
$$= 192.$$

Example 2. *Find the eighth term of the series*
$$2, -6, 18, -54, + . . .$$

For this series $r = -3$.

Using ar^{n-1},
$$\text{eighth term} = 2 \times (-3)^{8-1} = 2 \times (-3)^7$$
$$= -4374.$$

Example 3. *Find the fifth term of the series in which the first term is* 100 *and the common ratio* 0·63.

Using ar^{n-1}, if x = the fifth term

	No.	log.
	0·63	$\bar{1}$·7993
		4
		$\bar{1}$·1972
	100	2
	15·75	1·1972

$$x = 100 \times (0 \cdot 63)^4.$$
$$\therefore \ \log x = \log 100 + 4 \log 0 \cdot 63$$
$$= 1 \cdot 1972$$
$$= \log 15 \cdot 75.$$

∴ the fifth term is 15·75.

Example 4. *The third term of a G.P. is 4·5 and the ninth is 16·2. Find the common ratio.*

Using ar^{n-1},

$$\text{third term} = ar^2 = 4·5$$
$$\text{ninth term} = ar^8 = 16·2.$$

Dividing $\qquad ar^8 \div ar^2 = 16·2 \div 4·5.$

$\therefore \qquad\qquad\qquad r^6 = 16·2 \div 4·5.$

Taking logs $\qquad 6 \log r = \log 16·2 - \log 4·5$
$$= 0·5563.$$

$\therefore \qquad\qquad \log r = 0·5563 \div 6$

$$= 0·0927$$
$$= \log 1·238.$$

$\therefore \qquad\qquad\qquad r = 1·238.$

No.	log.
16·2	1·2095
4·5	0·6532
	0·5563

198. Geometric mean.

If three members are in G.P., the middle term is called the geometric mean of the other two.

Let a, b, c be three numbers in G.P.

Then by the definition of § 195

$$\frac{b}{a} = \frac{c}{b}.$$

$\therefore \qquad\qquad b^2 = ac,$

and $\qquad\qquad b = \sqrt{ac}.$

Exercise 55.

1. Write down the next three terms of each of the following series:

    (a) 4, 10, 25.
    (b) 16, 4, 1.
    (c) 16, — 24, 36.
    (d) 0·3, 0·03, 0·003.
    (e) 3, 0·45, 0·0675.

2. Find the seventh term of the series 5, 10, 20, . . .
3. Find the seventh term of the series 6, — 4, $2\frac{2}{3}$, . . .
4. Find the fifth term of the series 1·1, 1·21, 1·331, . . .
5. Find the sixth term of the series — 0·5, 0·15, — 0·045, . . .

6. Write down the $2n$th and the $(2n + 1)$th terms of the series:

    (1) $a, ar, ar^2, \ldots$
    (2) $a, -ar, ar^2, \ldots$

7. The first term of a G.P. is $1 \cdot 05$ and the sixth term is $1 \cdot 3401$. Find the common ratio.

8. The fifth term of a G.P. is $1 \cdot 2166$ and the seventh term is $1 \cdot 3159$. Find the common ratio.

9. Find the geometric mean in each of the following cases:

    (1) 3 and 5.
    (2) $4 \cdot 2$ and $3 \cdot 6$.

10. Insert two geometric means between 5 and $13 \cdot 72$.

11. A man was appointed to a post at a salary of £1000 a year with an increase each year of 10 per cent. of his salary for the previous year. How much does he receive during his fifth year?

12. The expenses of a company are £200 000 a year. It is decided that each year they shall be reduced by 5 per cent. of those for the preceding year. What will be the expenses during the fourth year, the first reduction taking place during the first year?

13. In a geometric series the first term is unity and the fifth term is $1 \cdot 170$ approx. Find the common ratio.

14. Insert three terms in geometric progression between 5 and 80.

---

199. The sum of n terms of a geometric series.

In addition to the symbols employed above,
Let S_n represent the sum of n terms of a G.P.

then $S_n = a + ar + ar^2 + \ldots + ar^{n-2} + ar^{n-1}$.   (1)

Multiply both sides by r,

then $rS_n = ar + ar^2 + ar^3 \ldots + ar^{n-1} + ar^n$.   (2)

Subtracting (1) from (2),

$$rS_n - S_n = ar^n - a$$

or $S_n(r - 1) = a(r^n - 1).$

$\therefore$
$$S_n = \frac{a(r^n - 1)}{r - 1} \quad \cdot \quad \cdot \quad \cdot \quad \text{(A)}$$

If (2) be subtracted from (1) above, the formula becomes:

$$S_n = \frac{a(1 - r^n)}{1 - r} \quad . \quad . \quad . \quad \textbf{(B)}$$

If $r > 1$ and positive, form (A) should be used.
If $r < 1$ or negative, form (B) should be used.

Note.—It will often be necessary to employ logarithms to evaluate r^n. They cannot of course be employed to evaluate the whole formula.

200. Worked examples.

Example 1. *Find the sum of seven terms of the series*

$$2, 3, 4\cdot5, \ldots$$
$$r = \tfrac{3}{2} = 1\cdot5.$$

Using $\qquad S = \dfrac{a(r^n - 1)}{r - 1}$,

and substituting $\quad S = \dfrac{2(1\cdot5^7 - 1)}{1\cdot5 - 1}$

logs may be used to find $1\cdot5^7$.

Let $\qquad x = 1\cdot5^7$.
then $\qquad \log x = 7 \log 1\cdot5$
$\qquad\qquad\qquad = 7 \times 0\cdot1761$
$\qquad\qquad\qquad = 1\cdot2327$
$\qquad\qquad\qquad = \log 17\cdot09.$
$\therefore \qquad 1\cdot5^7 = 17\cdot09 \text{ (approx.)}.$
$\therefore \qquad S = \dfrac{2(17\cdot09 - 1)}{0\cdot5} = 4 \times 16\cdot09$
$\qquad\qquad\qquad = 64\cdot36 = 64\cdot4 \text{ approx.}$

Example 2. *Find the sum of seven terms of the series*

$$4, -8, 16, \ldots$$
$$r = -2.$$

Using $\qquad S = \dfrac{a(1 - r^n)}{1 - r}$,

and substituting $\quad S = \dfrac{4\{1 - (-2)^7\}}{1 - (-2)} = \dfrac{4(1 + 128)}{3}$

$\qquad\qquad\qquad = \tfrac{4}{3} \times 129 = 172.$

Exercise 56.

1. Find the sums of the following series:

    (a) $1\cdot5, 3, 6, \ldots$ to six terms.
    (b) $30, -15, 7\frac{1}{2}, \ldots$ to eight terms.
    (c) $\frac{1}{2}, -\frac{1}{4}, \frac{1}{8}, \ldots$ to six terms.

2. Find the sum of the first six terms of the series

$$5, 2\cdot5, 1\cdot25, \ldots$$

3. Find the sum of the first six terms of the series

$$1 - \frac{2}{3} + \frac{4}{9}, \ldots$$

4. Find the sum of the first six terms of the series

$$1 + 1\cdot4 + 1\cdot96 + \ldots$$

5. Find the sum of the first twelve terms of the series

$$4 + 5 + 6\cdot25 + \ldots$$

6. If the first and third terms of a G.P. are 3 and 12, find the sum of eight terms.

7. If the third and fourth terms of a G.P. are $\frac{1}{9}$ and $\frac{1}{3}$, respectively, find the eighth term and the sum of eight terms.

8. Find the sum of $20 + 18 + 16\cdot2 + \ldots$ to six terms.

Infinite Geometric Series

201. Increasing series.

When the common ratio of a geometric series is numerically greater than unity, as in the series

$$1, 2, 4, 8, \ldots$$
$$2\cdot5, 7\cdot5, 22\cdot5, \ldots$$

the terms increase in magnitude. The sum of n terms increases as n increases. If the number of terms increases without limit—that is, n is greater then any number we may select, however great—then the sum of these terms will also increase without limit, *i.e.* it will become infinitely great, or, to use the mathematical term, approach " infinity ", which is denoted by the symbol ∞.

We may say, then, that as n, the number of terms, approaches infinity, S_n, the sum of these terms, also approaches infinity. This may be expressed by the following notation.

If $n \longrightarrow \infty$, then $S_n \longrightarrow \infty$.

202. Decreasing series.

If, however, the common ratio is numerically less than unity, as in the following series,

$$\tfrac{1}{3}, \tfrac{1}{6}, \tfrac{1}{12}, \ldots$$
$$0 \cdot 2,\ 0 \cdot 02,\ 0 \cdot 002,\ \ldots$$

then as the number of terms increases, the terms themselves decrease. Using the terms employed above, we may say that, as n increases without limit, the terms themselves decrease without limit, and ultimately become indefinitely small.

We cannot say, however, that the sum of these terms increases without limit, as n increases without limit. That is a matter for further investigation.

203. Recurring decimals.

There is an example, arising from Arithmetic, which will assist in coming to conclusions on this important question, viz. that of a recurring decimal. We know that

$$\tfrac{1}{9} = 0 \cdot 1111 \ldots$$

in which 1 recurs without limit.

The decimal is in effect the geometric series

$$\frac{1}{10} + \frac{1}{10^2} + \frac{1}{10^3} + \ldots$$

in which there is no limit to the number of terms. It is an example of what is called an **infinite series**. But we know that the sum of all these terms, no matter how many are taken, is ultimately equal to the finite fraction $\tfrac{1}{9}$.

If we find the sum of finite numbers of terms, we get:

$$S_1 = \frac{1}{10}.$$
$$S_2 = \frac{1}{10} + \frac{1}{10^2} = \frac{11}{10^2}.$$
$$S_3 = \frac{1}{10} + \frac{1}{10^2} + \frac{1}{10^3} = \frac{111}{10^3} \quad \text{and so on.}$$

The difference between $\frac{1}{9}$ and the sums of these is:

$$\frac{1}{9} - S_1 = \frac{1}{90}$$
$$\frac{1}{9} - S_2 = \frac{1}{900}$$
$$\frac{1}{9} - S_3 = \frac{1}{9000}$$

and in general, finding the sum of n terms by using the formula

$$S_n = \frac{a(1 - r^n)}{1 - r}$$

we get
$$S_n = \frac{\frac{1}{10}\left(1 - \frac{1}{10^n}\right)}{1 - \frac{1}{10}}$$

$$= \frac{1}{9}\left(1 - \frac{1}{10^n}\right) = \frac{1}{9} - \frac{1}{9 \times 10^n}.$$

Examining these results, it is seen that the difference between $\frac{1}{9}$ and the various sums, $S_1, S_2, S_3, \ldots S_n$ decreases as n increases. In general, the difference between $\frac{1}{9}$ and the sum of n terms is $\dfrac{1}{9 \times 10^n}$.

As n increases without limit, this difference decreases without limit—*i.e.*, it tends to become zero—and the sum approaches to equality with $\frac{1}{9}$. It can never be greater than $\frac{1}{9}$.

Using the previous notation, we can express the result thus

$$\text{as } n \longrightarrow \infty, \ S_n \longrightarrow \tfrac{1}{9}.$$

There is thus a limit to which S_n approaches and which it cannot exceed.

204. A geometrical illustration.

The approach of the sum of a geometric series to a limit may be illustrated by a graphical representation of the series

$$\frac{1}{2} + \frac{1}{4} + \frac{1}{8} + \frac{1}{16} + \cdots$$

or
$$\frac{1}{2} + \frac{1}{2^2} + \frac{1}{2^3} + \frac{1}{2^4} + \cdots$$

Let the rectangle *ABCD* (Fig. 52) represent a unit of area.

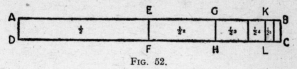

FIG. 52.

Let *E* be the midpoint of *AB* and draw *EF* perpendicular to *DC*.

Then rectangle *AEFD* represents $\frac{1}{2}$ of a unit.

Bisecting the rectangle *EBCF* by *GH*, then rectangle *EGHF* represents $\frac{1}{4}$ or $\frac{1}{2^2}$ of a unit.

Continuing the process of bisecting the rectangle left over after each bisection, we get a series of rectangles whose areas represent the terms of the above series. These rectangles diminish in area as we represent more and more terms of the series in this way.

The rectangle *AKLD* represents the sum of the four terms of the series

$$\frac{1}{2} + \frac{1}{2^2} + \frac{1}{2^3} + \frac{1}{2^4}.$$

As more divisions are made and the sum of more terms represented by a rectangle, this rectangle approaches nearer to the area of the whole rectangle—*i.e.*, 1—but can never exceed it. Consequently 1 is a limit which the sum of the series approaches as the number of terms is increased without limit, but which it can never exceed, no matter how many terms are taken.

If the series $\frac{1}{2} + \frac{1}{2^2} + \frac{1}{2^3} + \ldots$ be summed by using

the formula $S_n = \dfrac{a(1 - r^n)}{1 - r}$,

we get $S_n = \dfrac{\frac{1}{2}\{1 - (\frac{1}{2})^n\}}{1 - \frac{1}{2}} = \dfrac{\frac{1}{2}\{1 - (\frac{1}{2})^n\}}{\frac{1}{2}}$.

$\therefore \quad S_n = 1 - (\frac{1}{2})^n.$

Examining this result, we see that $(\frac{1}{2})^n$ decreases as *n* increases. If n be increased without limit then $(\frac{1}{2})^n$ decreases without limit—*i.e.*, it approaches zero.

$\therefore$ we can say that $S_n \longrightarrow 1$ as $n \longrightarrow \infty$.

205. The sum to infinity.

The above suggests the general treatment of this question.

Using
$$S_n = \frac{a(1 - r^n)}{1 - r}$$

i.e.,
$$S_n = \frac{a - ar^n}{1 - r}$$

we have
$$S_n = \frac{a}{1 - r} - a \cdot \frac{r^n}{1 - r}.$$

Considering the term $a \cdot \dfrac{r^n}{1-r}$, if r is a proper fraction (*i.e.*, it lies between $+1$ and -1) then r^n diminishes as n increases or, with the previous notation,

$$\text{as } n \longrightarrow \infty \, , \, r^n \longrightarrow 0 \text{ and } a \cdot \frac{r^n}{1 - r} \longrightarrow 0.$$

Thus the right hand side approaches $\dfrac{a}{1 - r}$ as a limit.

This is the "*limiting sum*" of the series and it is called the "*sum to infinity*".

If it be represented by S_∞

then
$$S_\infty = \frac{a}{1 - r}.$$

206. Worked examples.

Example 1. *Sum to infinity the series*
$$2 + \tfrac{1}{2} + \tfrac{1}{8} + \cdots$$

Here $a = 2$, $r = \frac{1}{4}$.

$\therefore$
$$S_\infty = \frac{2}{1 - \frac{1}{4}} = 2 \div \tfrac{3}{4}$$
$$= 2\tfrac{2}{3}.$$

Example 2. *Find the sum to infinity of the series*
$$5 - 1 + \tfrac{1}{5} - \cdots$$

Here $a = 5$, $r = -\frac{1}{5}$.

$$\therefore \qquad S_\infty = \frac{5}{1 - (-\frac{1}{5})} = 5 \div 1\frac{1}{5}$$
$$= 4\frac{1}{6}.$$

Exercise 57.

1. To what limits will the following series tend as the number of terms increases indefinitely?

    (a) $\frac{1}{2} + \frac{1}{8} + \frac{1}{32} + \ldots$
    (b) $\frac{1}{4} + \frac{1}{16} + \frac{1}{64} + \ldots$
    (c) $\frac{1}{2} + \frac{1}{4} + \frac{1}{8} + \frac{1}{16} + \frac{1}{32} + \ldots$

2. To what limits will the following series tend as the number of terms increases indefinitely?

    (a) $0 \cdot 1 + 0 \cdot 001 + 0 \cdot 000\,01 + \ldots$
    (b) $0 \cdot 06 + 0 \cdot 0006 + 0 \cdot 000\,006 + \ldots$
    (c) $0 \cdot 16 + 0 \cdot 0016 + 0 \cdot 000\,016 + \ldots$

What is the connection between these series?

3. To what limit does an infinite number of terms in the following series tend?

$$1 - \frac{1}{2} + \frac{1}{4} - \frac{1}{8} + \ldots$$

Show the connection with the series in question 1.

4. Show that the sum of n terms of the series

$$1 + \frac{1}{3} + \frac{1}{9} + \ldots$$

is
$$\frac{3}{2}\{1 - (\frac{1}{3})^n\}.$$

Hence show what limit this series approaches.

5. Find the sum of n terms of the series

$$1 + \frac{a}{b} + \frac{a^2}{b^2} \ldots$$

when
$$\frac{a}{b} < 1.$$

Hence find the limit approached by the series as the number of terms becomes infinitely great.

6. Find the limiting sum, or the sum to infinity of the following series :

(a) $\frac{8}{3} + \frac{4}{9} + \frac{2}{27} + \ldots$

(b) $5 - 1 + \frac{1}{5} - \ldots$

(c) $9 - 6 + 4 - \ldots$

7. Find the sum to infinity of the series:

(a) $1 + \dfrac{1}{1\cdot04} + \dfrac{1}{1\cdot04^2} + \ldots$

(b) $\sqrt{2} + 1 + \dfrac{1}{\sqrt{2}} + \ldots$

8. The sum to infinity of a series is 15, and the first term is 3. Find the common ratio.

9. A superball is dropped from a height of 10 m. At each rebound it rises to a height which is 0·9 of the height from which it has just fallen. What is the total distance through which the ball will have moved before it finally comes to rest?

10. The yearly output of a silver mine is found to be decreasing by 25 per cent. of its previous year's output. If in a certain year its output was £25 000 what could be reckoned as its total future output?

---

207. Simple and compound interest.

The accumulation of money when put to interest furnishes examples of arithmetic and geometric series. When money is put out at simple interest, the interest is payable for each year, but is not added to the principal.

For example the interest payable on £100 at 5 per cent. for 1, 2, 3 . . . years will be £5, £10, £15 . . ., these sums forming a series in A.P. and the **interest varies directly as the time.**

But if money is lent at compound interest, the interest is added each year to the principal, and for the following year the interest is calculated on their sum.

Suppose £1 to be invested at 5 per cent. compound interest. Then the interest for the first year is $£\frac{5}{100}$ or £0·05.

∴ the amount at the end of the year is £1·05.

and   ,,   ,,   ,,   ,,   of £P is £P × 1·05.

Consequently the ratio of the amount at the end of a year to that at the beginning is always $1·05$.

This corresponds to the ratio of a geometric series.

∴ amount at the end of the second year is

$$P \times 1·05 \times 1·05 = P \times 1·05^2.$$

∴ amount at the end of the third year is

$$P \times 1·05^2 \times 1·05 = P \times 10·5^3.$$

∴ amount at the end of the fourth year is

$$P \times 1·05^3 \times 1·05 = P \times 1·05^4.$$

∴ amount at the end of the nth year

$$= P \times 1·05^n.$$

These amounts at the end of successive years, viz.:

$$P \times 1·05, \ P \times 1·05^2, \ P \times 1·05^3 \ . \ . \ .$$

constitute a geometric series.

Let $M =$ the amount at the end of n years

then

$$M = PR^n.$$

In this formula, as we have seen in other cases, any one of the four quantities may be the subject of the formula.

Thus

$$P = \frac{M}{R^n}.$$

This enables us to find the sum of money which will produce £M in n years.

Again

$$R^n = \frac{M}{P}.$$

∴

$$R = \sqrt[n]{\frac{M}{P}},$$

from which the rate of interest can be discovered.

Again

$$n \log R = \log M - \log P.$$

∴

$$n = \frac{\log M - \log P}{\log R}$$

whence the time taken for P to amount to M is found.

208. Accumulated value of periodical payments.

Suppose that £P is invested each year for 10 years at 5 per cent. C.I., each investment being made at the beginning of a year.

Using the above formula:

The first £P at the end of 10 years amounts to
$$P \times 1 \cdot 05^{10}.$$

The second £P at the end of 9 years amounts to
$$P \times 1 \cdot 05^{9}.$$

The third £P at the end of 8 years amounts to
$$P \times 1 \cdot 05^{8}.$$

And finally, the last £P invested bears interest for 1 year and amounts to
$$P \times 1 \cdot 05.$$

Then the accumulated value of the investments amounts to
$$P \times 1 \cdot 05^{10} + P \times 1 \cdot 05^{9} + P \times 1 \cdot 05^{8} \ldots + P \times 1 \cdot 05,$$
or, reversing the series,
$$P \times 1 \cdot 05 + P \times 1 \cdot 05^{2} + P \times 1 \cdot 05^{3} \ldots + P \times 1 \cdot 05^{10}.$$

This is a geometric series and using the form
$$S_n = \frac{a(r^n - 1)}{r - 1}$$

the accumulated value of the investments is
$$\frac{1 \cdot 05(1 \cdot 05^{10} - 1)}{1 \cdot 05 - 1} \times P.$$

209. Annuities.

An annuity is a series of equal annual payments extending over a specified number of years, or for the life of the annuitant.

A ground rent is a similar financial transaction, the holder of the freehold receiving an annual payment, called ground rent, for the number of years specified in the lease. Ground rents and annuities are constantly being bought and sold, and the method of calculating the amount to be paid by the purchaser can be determined by means of the above results. This amount will depend upon the rate of interest which the purchaser expects to receive on his investment.

Let the rate of interest expected be 4 per cent.

The price is obtained by finding the present value of each of the payments, as follows:

From the formula $\qquad M = PR^n$

we get as shown $\qquad P = \dfrac{M}{R^n}$.

P is the amount which produces M in n years at the given rate per cent.; it is called the **present value of M** due in n years.

If £A be the annual payment and P be its present value, then for the first payment due in 1 year

$$P = \frac{A}{R} = \frac{A}{1 \cdot 04}$$

when the rate per cent. is 4.

For the second payment

$$P = \frac{A}{1 \cdot 04^2}.$$

For the third payment

$$P = \frac{A}{1 \cdot 04^3},$$

and so on.

$\therefore$ the total present value

$$= \frac{A}{1 \cdot 04} + \frac{A}{1 \cdot 04^2} + \frac{A}{1 \cdot 04^3} + \cdots$$

$$= A \left\{ \frac{1}{1 \cdot 04} + \frac{1}{1 \cdot 04^2} + \frac{1}{1 \cdot 04^3} + \cdots \right\}$$

$$= A \left\{ \frac{\dfrac{1}{1 \cdot 04} \left[1 - \left(\dfrac{1}{1 \cdot 04} \right)^n \right]}{1 - \dfrac{1}{1 \cdot 04}} \right\}$$

$$= \frac{A}{0 \cdot 04} \left\{ 1 - \frac{1}{1 \cdot 04^n} \right\}$$

This can then be evaluated for any value of A.

The terms of the above series decrease and if the ground rent is a perpetual one, or the lease is a very long one, the present value becomes the sum to infinity of the above series—*i.e.*,

$$\text{Present value} = A \left\{ \frac{\dfrac{1}{1 \cdot 04}}{1 - \dfrac{1}{1 \cdot 04}} \right\}$$

$$= A \left\{ \frac{\dfrac{1}{1 \cdot 04}}{\dfrac{1 \cdot 04 - 1}{1 \cdot 04}} \right\} = A \frac{1}{0 \cdot 04} = A \times \frac{100}{4} = A \times 25.$$

The ground rent is then said to be worth **25 years purchase**. It is always found by dividing 100 by the rate per cent.

Exercise 58.

1. If £100 be invested at the beginning of each year for 10 years at 3 per cent. C.I., find the accumulated value a year after the last amount is invested.

2. An annuity of £600 a year is allowed to accumulate at 3 per cent. C.I. for 8 years. What was the total amount at the end?

3. A man saves £25 every half-year and invests it at C.I. at $4\frac{1}{2}$ per cent. What will be the amount of his savings in 8 years if the last amount saved bears interest for 6 months?

4. Find the present value of an annuity of £300 for 10 years, reckoning C.I. at 4 per cent., the first payment being due one year after purchase.

5. What should be the purchase price of an annuity of £500 for 8 years, reckoning C.I. at $3\frac{1}{2}$ per cent.?

6. A man wished to endow in perpetuity an institution with a yearly sum of £200. If C.I. be reckoned at 4 per cent., what amount will be needed for it?

7. A man retires at 65, when the expectation of life is 10·34 years, with a pension of £200. What single payment would be the equivalent of this, reckoning C.I. at 4 per cent.?

8. A pension of £6000 per annum was awarded to Nelson and his heirs for ever. If this be commuted into a single payment, what should that be, reckoning C.I. at $2\frac{1}{2}$ per cent.?

APPENDIX

THE following brief statement of Permutations, combinations and the Binomial Theorem, and a note on the roots of a quadratic, equation are given for the benefit of students who may need to use them in the Differential Calculus or other branches of more advanced Mathematics.

Permutations and Combinations

I. Permutations.

Consider the following example:

A party of 6 people arrived at a theatre and obtained 4 seats together and 2 separate. In how many different ways could the 4 seats in a row be filled if there are no restrictions as to where any of the 6 may sit?

Consider the first seat. Since any one of the 6 people may sit in it, it can be filled in 6 different ways. With each of these 6 ways, the second seat can be filled in 5 different ways, since 5 people are left to choose from.

∴ *there are* (6 × 5) *different ways of filling the first two seats*.

With each of the 6 × 5 or 30 ways of filling the first two seats, there are 4 ways of filling the third seat, since 4 people are left to choose from.

∴ *there are* (6 × 5 × 4) *different ways of filling the first three seats*.

Similarly the fourth seat can be filled in 3 ways and

∴ *there are* (6 × 5 × 4 × 3) *different ways of filling the 4 seats—i.e.*, 360 *ways*.

Arrangements of a number of different objects in a row are called Permutations and the above problem was that of the permutation of 6 things 4 at a time.

This is expressed by a special notation—viz., $^6P_4$ or $_6P_4$.

It will be seen that if the people mentioned above had 6 seats together, the number of permutations or arrangements in these seats would be 6 × 5 × 4 × 3 × 2 × 1. This product of all the integral numbers from 1 to 6 inclusive is called Factorial 6 and is expressed by the ⌊6 or 6 !

In general, the product of the integral numbers from 1 to n inclusive is denoted by $\lfloor n$ or n !

Thus $\lfloor n = n(n-1)(n-2) \ldots 3, 2, 1$ and is called Factorial n.

2. Permutations of n things r at a time, or $^{n}P_{r}$.

This is the general treatment of the above special case and the method adopted to find the formula is the same.

There are r places to be filled and n different things to choose from.

The 1st place can be filled in n ways
  ,, 2nd   ,,   then be filled in $(n-1)$ ways

since with each of the n ways of filling the 1st place, each of the $(n-1)$ ways of filling the second can be associated.

∴ there are $n(n-1)$ ways of filling the first two places.

Similarly there are

$$n(n-1)(n-2) \text{ ways of filling the three places,}$$
and $n(n-1)(n-2)(n-3)$ ways of filling the four places.

∴ by inspection there are

$$n(n-1)(n-2)(n-3) \ldots \{n-(r-1)\} \text{ ways}$$
of filling the r places.

∴ $^{n}P_{r} = n(n-1)(n-2)(n-3) \ldots (n-r+1).$

If the n things are *all* arranged among themselves, then the last factor becomes $(n-n+1)$ or 1.

∴ $^{n}P_{n} = \lfloor n.$

3. Combinations.

The problem solved above—viz., the number of different ways of filling up 4 seats by 6 people—might have been approached in another way.

    (1) We could find the number of different *sets* or *groups* of 4 that could be formed from 6 people.

    (2) Each group could then be arranged in the seats in $\lfloor 4$ ways.

The product of these two numbers must give the total number of ways of filling the 4 seats—*i.e.*, the permutations of 6 things 4 at a time.   The difficulty at present is that of finding the number of groups.

Let $x =$ the number of groups
then, by the above reasoning,

$$x \times \lfloor 4 = {}^6P_4.$$

$$\therefore \qquad x = {}^6P_4 \div \lfloor 4.$$

Thus we can find the number of groups, when we know the number of permutations.

Such groups are called Combinations and a notation similar to that used for Permutation is employed to express them.

Thus the number of groups or combinations of 6 different things 4 at a time is denoted by ${}^6C_4$, and in general, the number of combinations of n things r at a time is denoted by ${}^nC_r$ or ${}_nC_r$.

4. The Combinations of n things r at a time.

With the same reasoning as that employed above we can deduce that

$$ {}^nC_r = {}^nP_r \div \lfloor r $$

or $\qquad {}^nC_r = \dfrac{n(n-1)(n-2) \ldots (n-r+1)}{\lfloor r}.$

Thus $\quad {}^nC_2 = \dfrac{n(n-1)}{\lfloor 2}$

$$ {}^nC_3 = \dfrac{n(n-1)(n-2)}{\lfloor 3} $$

$$ {}^{10}C_4 = \dfrac{10 \times 9 \times 8 \times 7}{1 \times 2 \times 3 \times 4} = 210 $$

$$ {}^8C_3 = \dfrac{8 \times 7 \times 6}{1 \times 2 \times 3} = 56. $$

The Binomial Theorem

5. Products of binomial factors.

It was shown in § 77 that

$$(x + a)(x + b) = x^2 + x(a + b) + ab.$$

Employing the methods used in the chapter we can show that $(x + a)(x + b)(x + c)$

$$= x^3 + x^2(a + b + c) + x(ab + bc + ca) + abc.$$

It should be noted that:

(1) The expression is arranged in **descending powers** of x.

(2) The **coefficients of these powers**, after the first, are the sets formed in every way of the letters a, b, c, (a) one at a time, (b) 2 at a time, (c) 3 at a time.

From the way in which this product is formed we can deduce the product of

$$(x + a)(x + b)(x + c)(x + d).$$

Arranging the powers of x in descending order, the coefficients of these powers will be

x^4—unity.

x^3—sum of the letters one at a time, *i.e.*,

$$(a + b + c + d).$$

x^2—sum of the letters two at a time, *i.e.*,

$$(ab + bc + ad + bd + cd + ac).$$

x—sum of the letters three at a time, *i.e.*,

$$(abc + bcd + acd + abd).$$

Term independent of x is $abcd$.

$\therefore$ the full product is

$$x^4 + x^3(a + b + c + d) + x^2(ab + bc + cd + ad + bd + ac)$$
$$+ x(abc + bcd + acd + abd) + abcd.$$

We have seen from § 4, that the number of ways of

(1) Grouping 4 letters 1 at a time is $^4C_1 = \dfrac{4}{1} = 4.$

(2)   ,,   4   ,,   2   ,,   ,, $^4C_2 = \dfrac{4 \cdot 3}{1 \cdot 2} = 6.$

(3)   ,,   4   ,,   3   ,,   ,, $^4C_3 = \dfrac{4 \cdot 3 \cdot 2}{1 \cdot 2 \cdot 3} = 4.$

(4)   ,,   4   ,,   4   ,,   ,, $^4C_4 = 1.$

In the above factors let $b = c = d = a$
then the left side is $(x + a)^4$.

In the expansion of it :

The coefficient of x^3 is $(a + a + a + a) = 4a$.

,,   ,,   ,, x^2 ,, $6a^2$.

,,   ,,   ,, x ,, $4a^3$.

The last term is a^4.

$$\therefore \quad (x + a)^4 = x^4 + 4x^3a + 6x^2a^2 + 4xa^3 + a^4$$

or $\quad (x + a)^4 = x^4 + {}^4C_1x^3a + {}^4C_2x^2a^2 + {}^4C_3xa^3 + a^4.$

By a similar process we can obtain the expansion

$$(x + a)^5 = x^5 + {}^5C_1x^4a + {}^5C_2x^3a^2 + {}^5C_3x^2a^3 + {}^5C_4xa^4 + a^5.$$

From a consideration of these results we can deduce the general case—viz.,

$$(x + a)^n = x^n + {}^nC_1x^{n-1}a + {}^nC_2x^{n-2}a^2 + {}^nC_3x^{n-3}a^3 + \quad \ldots + a^n$$

or

$$(x + a)^n = x^n + nx^{n-1}a + \frac{n(n-1)}{\lfloor 2} x^{n-2}a^2$$
$$+ \frac{n(n-1)(n-2)}{\lfloor 3} x^{n-3}a^3 + \ldots + a^n.$$

This is called the **Binomial Theorem.**

The above reasoning is independent of the values of x and a. It will therefore hold if a be replaced by $-a$. Then

$$(x - a)^n = x^n + nx^{n-1}(-a) + \frac{n(n-1)}{1 \cdot 2} x^{n-2}(-a)^2 + \quad \ldots + (-a)^n.$$

Since odd powers of $(-a)$ are negative and even powers are positive, the terms will be alternately $+$ve and $-$ve.

$$\therefore (x - a)^n = x^n - nx^{n-1}a + \frac{n(n-1)}{1 \cdot 2} \cdot x^{n-2}a^2$$
$$- \frac{n(n-1)(n-2)}{1 \cdot 2 \cdot 3} x^{n-3}a^3 + \ldots + (-a)^n.$$

In the above results let $x = 1$.

Then

$$(1 + a)^n = 1 + na + \frac{n(n-1)}{1 \cdot 2} a^2 + \frac{n(n-1)(n-2)}{1 \cdot 2 \cdot 3} a^3 + \\ \ldots + a^n$$

and

$$(1 - a)^n = 1 - na + \frac{n(n-1)}{1 \cdot 2} a^2 - \frac{n(n-1)(n-2)}{1 \cdot 2 \cdot 3} a^3 + \\ \ldots + (-a)^n.$$

Every binomial expression can be reduced to one of these forms as follows:

$$(x + a)^n = \left\{ x\left(1 + \frac{a}{x}\right) \right\}^n = x^n\left(1 + \frac{a}{x}\right)^n$$

$$= x^n\left\{ 1 + n \cdot \frac{a}{x} + \frac{n(n-1)}{1 \cdot 2} \left(\frac{a}{x}\right)^2 + \ldots + \left(\frac{a}{x}\right)^n \right\}$$

Similarly,

$$(x - a)^n = x^n\left(1 - \frac{a}{x}\right)^n.$$

A complete proof of the Binomial Theorem requires a more advanced knowledge of Algebra than is provided in this book. The demonstration given above assumes that n is a positive integer. In using the theorem later the important question will arise:

Is the Binomial Theorem true for fractional and negative indices?

It can be proved that the *form* holds for all values of n. For example of $n = \frac{1}{2}$.

$$(1 + x)^{\frac{1}{2}} = 1 + \frac{1}{2}x + \frac{\frac{1}{2}(\frac{1}{2} - 1)}{1 \cdot 2} x^2 + \frac{\frac{1}{2}(\frac{1}{2} - 1)(\frac{1}{2} - 2)}{1 \cdot 2 \cdot 3} x^3 + \ldots$$

When n is a positive integer, a term in the series will ultimately be reached when one of the factors

$$n(n - 1)(n - 2) \ldots$$

will become $(n - n)$ and will vanish, as will all succeeding products.

The number of terms therefore is finite and will clearly be $n + 1$.

Consequently when a value is assigned to x in $(1 + x)^n$ the sum of the series will be a finite number and the series is said to be **convergent**.

But if n is fractional or negative, none of the factors $n(n - 1)(n - 2)$ . . . will vanish, and the number of terms is **infinite**. If the sum increases without limit as n increases the series is said to be **divergent**. But it can be shown that if x, in $(1 + x)^n$, is numerically less than unity, the sum of the series will approach a limit, as is the case with certain geometric series (see § 205). Subject to this condition, the series is always **convergent**.

The Roots of a Quadratic Equation

Writing the general quadratic equation

$$ax^2 + bx + c = 0$$

in the form $\qquad x^2 + \dfrac{b}{a}x + \dfrac{c}{a} = 0 \qquad . \quad . \quad . \quad . \quad$ (A)

the solution is $\qquad x = \dfrac{-b \pm \sqrt{b^2 - 4ac}}{2a}$. (§ 127)

Let α and β be the two roots

then $\alpha + \beta = \dfrac{-b + \sqrt{b^2 - 4ac}}{2a} + \dfrac{-b - \sqrt{b^2 - 4ac}}{2a}$

$$= -\frac{2b}{2a} = -\frac{b}{a}.$$

= coeff. of x in equation (A).

Again $\alpha\beta = \dfrac{-b + \sqrt{b^2 - 4ac}}{2a} \times \dfrac{-b - \sqrt{b^2 - 4ac}}{2a}$

$$= \frac{(-b)^2 - (\sqrt{b^2 - 4ac})^2}{4a^2} \qquad \text{(§ 83)}$$

$$= \frac{c}{a}$$

= term independent of x in (A).

Summarising, $\qquad \alpha + \beta = -\dfrac{b}{a}$

$$\alpha\beta = \frac{c}{a} \quad \text{(cf. § 77)}$$

Nature of the roots of a quadratic equation.

If $(b^2 - 4ac)$ is negative, the square root has no arithmetical meaning. It is customary to speak of such a root as **imaginary**, while the square root of a positive number is called **real**.

Hence, for the roots of a quadratic,

(1) If $b^2 > 4ac$, the roots are **real** and different.
(2) If $b^2 = 4ac$,, ,, ,, **equal**.
(3) If $b^2 < 4ac$,, ,, ,, **imaginary**.

Such a number as $\sqrt{-p}$ can be written

$$\sqrt{p \times (-1)} = \sqrt{p} \times \sqrt{-1}.$$

The number $\sqrt{-1}$ is usually denoted by i.

Hence, as an example, $\sqrt{-9}$ can be written $\pm \sqrt{9 \times -1}$
$$= \pm 3i.$$

Example.—The roots of the equation $x^2 + 2x + 5 = 0$ are given by

$$x = \frac{-2 \pm \sqrt{4 - 20}}{2}$$

$$= \frac{-2 \pm \sqrt{-16}}{2}$$

$$= \frac{-2 \pm 4i}{2}$$

$$= -1 \pm 2i.$$

Note.—It will be found that the graph of $x^2 + 2x + 5$ does not cut the x axis, *i.e.*, it does not equal zero for any real value of x.

ANSWERS

Exercise 1.

1. (1) $100x$.   (2) $\dfrac{n}{100}$.

2. $\dfrac{100a}{n}$.

3. $100an + mb$.

4. (1) $1000a$.   (2) $3x$.   (3) $\dfrac{y}{1000}$.

5. $28 - n$.   6. $x + 50$, or $50 - x$.   7. $\dfrac{a}{x}$.

8. (1) x m.   (2) $\dfrac{180\,000}{x}$.

9. $2n + 5$.   10. (1) $x + 2$, $x - 2$.   (2) $x + 1$, $x - 1$.

11. $\dfrac{xa + yb}{100}$.   12. $\dfrac{2x + 5}{6y}$.

13. $a - b$.   14. xy.

15. $a + b$.   16. $\dfrac{x + 2}{y - 5}$.

17. $mv + nu$.   18. $100a + b$.

19. xv; $\dfrac{y}{v}$.

20. The original numbers are the digits of the final result, m and n.

Exercise 2.

1. (a) 10 doz.; 120.   (b) $10a$; 120.
2. (a) $5 \times 73 = 365$.   (b) $5b = 365$.
3. $3a + 4b$; 47.   4. $19a$; 47·5.
5. (1) $26b$.   (2) $19x$.   (3) $12a$.   (4) $4x$.
6. (1) $2a + 4b$.   (2) $12p + 3q$.   (3) $6a + 3b + 4$.
7. (1) $10a + 2b$.   (2) $2b + c - d$.   (3) $8x + 2z$.
8. (1) 13.   (2) 7.   (3) 13.   (4) 11.
9. (1) $8ab = 64$.   (2) $3ax + bx = 42$.
    (3) $3xy = 45$.   (4) $4ab + 4bx - ay = 36$.
10. $3\frac{3}{4}$.   11. 22.
12. (1) 66.   (2) 29.   (3) 30.
13. $3n + 12$.   14. $15a$.
15. a, $a + d$, $a + 2d$, $a + 3d$, $a + 4d$; $5a + 10d$.
16. $2n + 5$, $2n + 8$, $2n + 11$, $2n + 14$, $2n + 17$; $10n + 55$.

p. 31.   Exercise 3.

1. $12a$.   2. $10xy$.   3. $2xy$.

4. $21mn$.   5. $\dfrac{ab}{12}$.   6. $4ab$.

7. $60abc$.   8. $\dfrac{xyz}{24}$.   9. x^3.

10. a^4.   11. x^4.   12. a^6.
13. $2a^5$.   14. $6x^7$.   15. $2a^2b^2$.
16. $6b^4$.   17. x^3y^3.   18. $7x^6a^2$.
19. $6a^6$.   20. $27a^6b^3$.   21. x^6.
22. $8a^{12}$.   23. $16a^{12}$.   24. $64a^4$.
25. 54.   26. 21.   27. 8.
28. 172.   29. 33.   30. 1.
31. 216.   32. 576.

p. 33.   Exercise 4.

1. a^2.   2. a^2.   3. $3x$.   4. $\dfrac{b^4}{2}$.

5. $2a^4$.   6. $5y^3$.   7. $3x^3$.   8. $2c^2$.
9. ab.   10. x^3y.   11. $5a^2b$.   12. $3xy$.
13. 2.   14. $\frac{5}{4}$.

p. 37.   Exercise 5.

1. $\dfrac{12a}{35}$.   2. $\dfrac{x}{15}$.

3. $\dfrac{4y + 5x}{xy}$.   4. $\dfrac{19a}{6b}$.

5. $\dfrac{11x}{40y}$.   6. $\dfrac{31x}{18y}$.

7. $\dfrac{20a^3 + 9b^2c}{24a^2bc^2}$. 8. $\dfrac{3x^2 + 4}{3x}$.

9. $\dfrac{6ab^2 + 5}{2b}$.   10. $\dfrac{40y^2 - 6}{5y^2}$.

11. $\dfrac{a^2 + a + 1}{a}$.   12. $\dfrac{x^2 - x + 1}{x}$.

13. $\dfrac{c + 2a + 3b}{abc}$.   14. $\dfrac{5bc - 4ac + 2ab}{a^2b^2c^2}$.

---

15. x^2y.   16. $\dfrac{4b}{a^2}$.   17. $\dfrac{3b^2}{2a^2c}$.

18. 1.   19. xy.   20. $\dfrac{a}{bc}$.

21. $\dfrac{4}{3a}$.

22. $\dfrac{3y}{2}$.

23 $\dfrac{9}{5q}$.

24. $\dfrac{2xy}{3}$.

25. $\dfrac{2x^2}{y}$.

26. $2y$.

27. 1.

28. $\dfrac{x^2}{y^2}$.

29. $\frac{2}{3}$.

30. $\dfrac{2x^4}{3y^2}$.

31. $\dfrac{x^2}{y^2}$.

32 $\dfrac{2a^3}{3b^2}$.

33. $\dfrac{8a^2}{3c}$.

34. $\dfrac{3y^2}{4}$.

35. $\dfrac{20}{xy}$.

36. $\dfrac{4a^2b^2}{c^4}$.

p. 43.   Exercise 6.

1. $15x + 18z$.   2. $6a^2 + 8ab$.
3. $18a^3 + 42a^2b - 36a^2c$.   4. $8x + y$.
5. $5x^3 + 4x^2$.   6. x.   7. $8x + 5y - 4z$.
8. $x - 2y - z$.   9. $2x - y + 2z$.   10. $a + 7b$.
11. $a - b$.   12. $a + b$.   13. $4x + 2y - 2z$.
14. $a + 5b - 5c$.   15. $3y$.   16. $a^2 + b^2$.
17. $x^3 + x^2y - x^3y + xy^3$.   18. $x^2 + 9x + 23$.
19. $9p^2 + 3pq + 15q^2$.   20. $5x^2y^2 - 3xy + 6x^2$.
21. $2x^4 + 8x^3$.

p. 45.   Exercise 7.

1. $6a - 9$.   2. $3a + 30b$.   3. $\frac{3}{2}$.
4. $3a^2 + 2ab - 2ac$.   5. $2p^2 + p$.
6. $x^2 + 3xy + 12y^2$.   7. $7bc - 2b^2 + 2c^2$.
8. $13x + 5$.   9. $34 - 6a + 4b$.   10. $x + 3y$.
11. $2a - 10b$.   12. $\dfrac{3x^2}{2}$.   13. $\dfrac{19c}{8}$.
14. $2a - (b - c)$.   15. $x - (y + z)$.   16. $2(a + 2b - 3c)$.
17. $x^2 - y(x - y)$.   18. $2a - 4b + 7c$.   19. $x + 2y - 2z$.
20. (1) 111.   (2) 6.

p. 53.   Exercise 8.

1. $0 + 4 - 2 + 4 - 6 = 0$.
2. $+ 8 + 2 - 14 + 4 - 6 = - 6°$ C.
3. (1) $11°$ C.   (2) $15°$ C.   (3) $30°$ C.
4. (1) $+ 5$.   (2) $- 5$.   (3) $+ 8$.
5. (1) $(a) + 2$. $(b) + 4$.   (2) $+ 5$.
6. (1) $(+ 4)$.   (2) $(+ 8)$.   (3) $(- 8)$.
    (4) $(- 4)$.   (5) $(+ 3)$.   (6) $(- 8)$.
    (7) 0.   (8) $(- 8)$.

7. (1) $7a$.   (2) $-7x$.   (3) $10ab$.
   (4) $-x - 8y$.   (5) $a - 7b$.   (6) $-x + 2y$.
   (7) $7x - 3y$.   (8) -8.
8. (1) $2x - 2y$.   (2) $2x - y - 7z$.
9. (1) $3a - (-5a)$.   (2) $5x - (6x)$.
   (3) $-3a - (-10a)$.
10. (1) $(-a)$.   (2) $(+a)$.   (3) $(-a)$.

p. 57.   Exercise 9.

1. (1) $(+36)$.   (2) (-36).   (3) (-36).
   (4) $(+36)$.   (5) $(+4)$.   (6) (-4).
   (7) (-4).   (8) $(+4)$.
2. (1) $(-a^2)$.   (2) $(+a^2)$.   (3) (-1).
   (4) $(+1)$.
3. (1) $(-4ab)$.   (2) $(+4ab)$.   (3) $(-20xy)$.
   (4) $(-20xy)$.   (5) $\left(-\dfrac{5x}{y}\right)$.   (6) $\left(+\dfrac{5x}{y}\right)$.

4. (1) $+24$.   (2) $+6a^3$.
   (3) $+3y$.   (4) $-6ab$.
5. (1) $-10x^3$.   (2) $a^2 - b^2$.
   (3) $-2ab^2$.   (4) $-a^2 + 2ab + ao$.
6. (1) $+20ab^2$.   (2) $6x^2 + 8x$.
   (3) $-2xz + 2yz$.
7. (1) $a^2, -a^3, a^4, -a^5$.
   (2) $4x^2, -8x^3, +16x^4, -32x^5$.
   (3) $\dfrac{b^2}{9}, -\dfrac{b^3}{27}, +\dfrac{b^4}{81}, -\dfrac{b^5}{243}$.
8. $\pm 9, \pm 3x^2, -x, -2a^2$.
9. (1) $+16x$.   (2) $5x$.   (3) $2y$.
   (4) $-2b$.   (5) $-2t$.   (6) $2x$.
   (7) $8x^3$.   (8) $-3x$.   (9) $-4ab$.
   (10) $6a^2c$.

10. (1) $-b$.   (2) $-x$.   (3) $-\dfrac{9x}{4}$.

p. 64.   Exercise 10.

1. (a) $2\frac{1}{2}$.   (b) 20.
2. (a) 16.   (b) 60.
3. (a) -16.   (b) -10.
4. (a) 80.   (b) -72.
5. (a) $12\cdot6$.   (b) $-0\cdot08$.
6. (a) $3\frac{1}{5}$.   (b) 5.
7. (a) $2\cdot5$.   (b) 12.
8. (a) 8.   (b) $\frac{4}{5}$.
9. (a) $\frac{8}{5}$.   (b) 16.
10. (a) $1\cdot8$.   (b) -45.

11. (a) 14·8. (b) 18¾.
12. (a) 10/9. (b) ⅓
13. 22. 14. 5. 15. 8. 16. 29.
17. − ½. 18. − 3. 19. 18⅘.
20. (a) 3.3/5 (b) 27½.
21. (a) − 80. (b) − 10.
22. − 35. 23. 8. 24. − 7. 25. 7/3.
26. 3. 27. 7/4. 28. − 5.5/2. 29. 2·7.
30. 2·9. 31. 3. 32. 4.5/19. 33. 36.
34. 16½.

p. 67. Exercise 11.

1. 12. 2. 72·5. 3. 42.
4. 15. 5. 13·5. 6. 21, 23, 25.
7. 10 km. 8. 40.
9. 11·9 cm, 10·1 cm. 10. 4 men; 60p.
11. 5. 12. Son 8 years, father 32 years.
13. 30. 14. 80 at 18p, 40 at 12p.
15. 23

p. 70. Exercise 12.

1. 175. 2. 154. 3. 113½.
4. 85·6. 5. 0·616. 6. 57⅘.
7. 4½. 8. 204. 9. ⅝.
10. 1390 (approx.). 11. 31·4.

p. 74. Exercise 13.

1. $r = \sqrt{\dfrac{A}{\pi}}.$ 2. $r = \sqrt[3]{\dfrac{3V}{4\pi}}.$

3. $r = \sqrt{\dfrac{3V}{\pi h}}.$ 4. $C = \dfrac{825H}{E}$

5. $B = \sqrt{\dfrac{F \times 112 \times 10^5}{A}}.$ 6. $l = \sqrt[3]{\dfrac{48EId}{W}}.$

7. $s = \dfrac{v^2 - u^2}{2a};\ 17\frac{1}{2}.$

8. (1) $V = IR.$ (2) $R = \dfrac{V}{I};\ 0·1$

9. $n = \sqrt{\dfrac{NR - 1}{r}};\ \frac{70}{9}.$ 10. $k = 1008;\ v = \dfrac{1008}{p}.$

p. 77. Exercise 14.

1. 8a. 2. 2a. 3. p.
4. 4b. 5. $\dfrac{3a - 2b}{a}.$ 6. $\dfrac{bp + c}{b}.$

7. $\dfrac{2(b - a)}{b}$. 8. $\dfrac{6 - 19b}{3a}$. 9. p.

10. $\dfrac{2b - 12}{3a}$. 11. $\dfrac{3b}{a - b}$. 12. $\dfrac{12a}{4 + 3a}$.

13. $\dfrac{a^2 + b^2}{a - b}$. 14. $\dfrac{m}{n - 3}$.

p. 82. Exercise 15.

1. $x = 3, y = 6$.
2. $x = 5, y = 8$.
3. $x = 12, y = 3$.
4. $x = 8, y = 3$.
5. $x = 3, y = 1$.
6. $x = 4, y = 5$.
7. $x = 7, y = 4$.
8. $x = 10, y = 7$.
9. $x = 6, y = 5$.
10. $x = 3, y = 10$.
11. $x = 6, y = 10$.
12. $a = 5, b = -\frac{3}{2}$.
13. $a = -2, b = 3$.
14. $p = -2, q = 4$.
15. $a = \frac{5}{2}, y = -2$.
16. $x = 9, y = 13$.
17. $a = -7, b = \frac{3}{2}$.
18. $x = 6, y = -4$.
19. $x = 8, y = 12$.
20. $x = 1\cdot35, y = 2\cdot7$.
21. $x = 4, y = 2$.
22. $P = 1\cdot8, Q = 0\cdot32$.
23. $x = \frac{15}{2}, y = -3$.
24. $x = 12, y = 6$.

p. 86. Exercise 16.

1. $x = 12, y = 10$.
2. 16, 12.
3. 20, 7.
4. 10, 6.
5. $m = 2, b = -3$. Equation is $y = 2x - 3$; when $x = 6$, $y = 9$.
6. $m = 20, b = 10$.
7. $a = 0\cdot5, b = 0\cdot6$.
8. $a = -1\cdot36, b = 1\cdot38$.
9. 10 m, 6 m.
10. £3·75; 2p.
11. Tie 49p; socks 24p.
12. $u = 10, a = 6, 125$ m.

p. 96. Exercise 17.

2. 10·4° C. 3. Males: (1) 33·9. (2) 18·1.
Females: (1) 36·8. (2) 20·2.

4. (1) 44·2 kg. (2) 65 kg. (3) 1·5 kg.
5. Accurate deductions simply on the basis of the graphs (which are anyway not regular) are certain to be un-realistic. However . . .
(1) 1978–9, (2) 126 000. (3) 1978–9.

6. 25·1 m/s.
7. $P = £13\cdot30$ is wrong and should be £15. £12·80; £2·50; £36.50.
8. About 360 kilotherms.
9.

x	−3	−2	−1	0	1	2	3	4	5
y	6	2·5	0	−1·5	−2	−1·5	0	2·5	6

3·45; −1·45.

p. 105. **Exercise 18.**

1. (1) 2.   (2) — 2.   2. (1) — 1.   (2) $\frac{2}{3}$.
3. (1) — $\frac{5}{4}$.   (2) $\frac{5}{6}$.   4. (1) 3.   (2) 2.
5. (1) 4.   (2) 6.   6. (1) — $\frac{1}{2}$.   (2) 1.
7. $a = 2, b = 3$. Equation is $y = 2x + 3$. Intercept is 3.
8. $x + 2y = 4$.

p. 108. **Exercise 19.**

1. A, (4, 4); B, (4·6, 1·2); C, (— 2, 3);
   D, (4, — 2); E, (— 1·4, — 3·4); F, (3, 0);
   G, (0, — 3).
2. (1·12, 1·12).   4. (2, 2).
5. They lie on a straight line parallel to OX.
6. They appear to lie on a straight line which passes through
   the origin.
7. For every point on it the x co-ordinate is + 3.

p. 116. **Exercise 20.**

1. All lines pass through the origin with different slopes.
2. All lines are parallel with slope 45°.
3. All lines are parallel with gradient $\frac{1}{2}$.
4. All lines have the same intercept on the y axis, viz. 2.
5. Intercepts are :   (1) 1·5 and 3.   (2) 4 and —2.
                        (3) 2 and 5.   (4) 2·5 and — 2.
6. (1) $x = 4, y = 3$.   (2) $x = 3, y = 1$.
   (3) $x = — 2, y = 2$.
7. $a = 3; — 1$.   8. $b = 1; 1$.
9. $a = 3, b = — 2$. Equation is $y = 3x — 2; — 2$.
10. $y = 2x + 3$.
11. $3y — 2x = 7$. Gradient $= \frac{2}{3}$.
12. $y — x = 2$. Gradient = 1.

p. 121. **Exercise 21.**

1. $ab + ay + bx + xy$.   2. $ce + cf + de + df$.
3. $acxy + adx + bcy + bd$.   4. $ab — ay — bx + xy$.
5. $ax — bx — ay + by$.   6. $ab + ay — bx — xy$.
7. $ab — ay + bx — xy$.   8. $ab + 3a + 2b + 6$.
9. $ab — 3a — 2b + 6$.   10. $ab + 3a — 2b — 6$.
11. $ab — 3a + 2b — 6$.   12. $x^2 + 12x + 35$.
13. $a^2b^2 + 9ab + 18$.   14. $x^2 + 13x + 30$.
15. $x^2 — 13x + 30$.   16. $x^2 + 7x — 30$.
17. $x^2 — 7x — 30$.   18. $p^2 — 4p — 96$.
19. $x^2 — 12xy + 32y^2$.   20. $x^2 + 4xy — 32y^2$.
21. $x^2 — 4xy — 32y^2$.   22. $2a^2 + 9ab + 10b^2$.
23. $9x^2 — 27xy + 20y^2$.   24. $28x^2 + 15x + 2$.
25. $6x^2 — 11x + 3$.   26. $9x^2 — 9x — 4$.

27. $1 - y - 12y^2$.
28. $18x^2 - 27x - 5$.
29. $14x^2 + 29xy - 15y^2$.
30. $18a^2 - 57ab + 35b^2$.
31. $a + b$.
32. $a + b - c$.
33. 12.
34. $5y - 5x = 0·5$.
35. (a) $x^3 - y^3$.   (b) $a^3 + 8$.
      (c) $1 + x^3$.   (d) $x^3 + 3x^2a + 3xa^2 + a^3$.

p. 124.   Exercise 22.

1. $x^2 + 4x + 4$.
2. $x^2 - 4x + 4$.
3. $a^2 + 6ab + 9b^2$.
4. $a^2 - 6ab + 9b^2$.
5. $4x^2 + 4xy + y^2$.
6. $x^2 - 4xy + 4y^2$.
7. $a^2b^2 + 20ab + 100$.
8. $x^2y^2 - 6xy + 9$.
9. $16x^2 + 40xy + 25y^2$.
10. $16x^2 - 40xy + 25y^2$.
11. $25x^2y^2 + 60xy + 36$.
12. $1 - 20x^2 + 100x^4$.
13. $25x^4 + 30x^2y^2 + 9y^4$.
14. $9x^2y^2 - 12xy^3 + 4y^4$.
15. $x^2 + \dfrac{2x}{y} + \dfrac{1}{y^2}$.
16. $\dfrac{1}{x^2} - \dfrac{2}{xy} + \dfrac{1}{y^2}$.
17. $a^2 + \dfrac{2a}{3} + \dfrac{1}{9}$.
18. $\tfrac{1}{4} - \dfrac{3y}{4} + \dfrac{9y^2}{16}$.
19. $x^2 + 2xy + y^2 + 2x + 2y + 1$.
20. $1 - 2x + 4y + x^2 - 4xy + 4y^2$.
21. $a^2 + b^2 + c^2 + 2ab - 2ac - 2bc$.
22. $x^2 + y^2 + z^2 - 2xy + 2xz - 2yz$.
23. $4x^2 + 9y^2 + 25z^2 + 12xy - 20xz - 30yz$.
24. $16a^2 + 4b^2 + 1 - 16ab - 8a + 4b$.
25. $x^3 + 3x^2y + 3xy^2 + y^3$.
26. $x^3 - 3x^2y + 3xy^2 - y^3$.
27. $a^3 + 6a^2 + 12a + 8$.
28. $a^3 - 6a^2 + 12a - 8$.
29. $p^3 + 3p^2q + 3pq^2 + q^3$.
30. $p^3 - 3p^2q + 3pq^2 - q^3$.
31. $8x^3 + 12x^2y + 6xy^2 + y^3$.
32. $x^3 - 6x^2y + 12xy^2 - 8y^3$.
33. $27a^3 - 27a^2 + 9a - 1$.
34. $1 - 9b + 27b^2 - 27b^3$.
35. $2xy$.
36. (1) $4ab$; (2) $-4ab$.
37. $40x$.
38. $9y^2 + 18y + 9$; 36.
39. $4a(x + a)$; 256 m^2.
40. $2a(x + y + 2a)$.

p. 126.   Exercise 23.

1. $a^2 - x^2$.
2. $p^2 - q^2$.
3. $a^2 - 4b^2$.
4. $16x^2 - 9$.
5. $4x^2 - 1$.
6. $1 - 36x^2$.
7. $1 - a^4$.
8. $4x^4 - 1$.
9. $x^4 - y^4$.
10. $9x^2y^2 - 4$.
11. $144x^2y^2 - 1$.
12. $\tfrac{1}{16}x^2 - 49$.
13. $(x + y)^2 - z^2$.
14. $(a + x)^2 - y^2$.
15. $(2a + 3b)^2 - 1$.
16. $(x - 2y)^2 - 36$.
17. $a^2 - 4(b + c)^2$.
18. $4x^2 - 9(y + z)^2$.
19. $x^2 - \tfrac{4}{9}$.
20. $\dfrac{x^2}{4} - \dfrac{y^2}{9}$.
21. $(x + a)^2 - \tfrac{1}{4}$.

p. 128.   Exercise 24.

1. $6(x + 2)$.
2. $a(3b + 2)$.
3. $2y(2x + y)$.
4. $2a(3a - 2b)$.
5. $7xy(2xy - 1)$.
6. $16(1 - 2a^2)$.
7. $a(a - b + c)$.
8. $x(x^2 + 3x - 1)$.
9. $a^2(15a - 5b + 3b^2)$.
10. $3ac(2a - 5c)$.
11. $ab(a + b - c)$.
12. $\dfrac{bc}{3}\left(\dfrac{c}{2} - \dfrac{a}{3}\right)$.
13. $7\cdot4(13^2 + a^2)$.
14. $18\cdot6(18\cdot6 + 1\cdot4) = 18\cdot6 \times 20 = 372$.

p. 131.   Exercise 25.

1. $(a + b)(x + y)$.
2. $(p + q)(c + d)$.
3. $(a - d)(b + e)$.
4. $(x - y)(a - c)$.
5. $(x + p)(x + q)$.
6. $(x - g)(x - h)$.
7. $(a + 5)(b + 6)$.
8. $(a - 5)(b - 6)$.
9. $(a - 5)(b + 6)$.
10. $(2a + 3)(b - 5)$.
11. $(ax - b)(x + a)$.
12. $(x - b)(x + a)$.

p. 132.   Exercise 26.

1. $(x + 2)(x + 1)$.
2. $(x - 2)(x - 1)$.
3. $(x + 3)(x + 2)$.
4. $(x - 3)(x - 2)$.
5. $(x + 6)(x + 1)$.
6. $(x + 5)(x + 4)$.
7. $(x - 10y)(x - 2y)$.
8. $(a - 12b)(a - 3b)$.
9. $(xy + 6)(xy + 9)$.
10. $(ab - 16)(ab - 3)$.
11. $(y - 9)(y - 12)$.
12. $(x - 7y)(x - 5y)$.
13. $(x - 2)(x + 1)$.
14. $(x + 2)(x - 1)$.
15. $(x + 3y)(x - 2y)$.
16. $(x - 3y)(x + 2y)$.
17. $(b - 3)(b + 1)$.
18. $(b + 3)(b - 1)$.
19. $(x + 16)(x - 3)$.
20. $(x - 16)(x + 3)$.
21. $(x - 11y)(x + 10y)$.
22. $(a - 12)(a + 1)$.
23. $(a - 4)(a + 3)$.
24. $(p + 9)(p - 8)$.
25. $(p - 36)(p + 2)$.
26. $(1 - 5x)(1 - 4x)$.
27. $(1 - 10x)(1 + 2x)$.
28. $(xy - 11)(xy + 8)$.
29. $(p + 9)(p - 5)$.
30. $(p - 7q)(p + 8q)$.

p. 134.   Exercise 27.

1. $(3x + 4)(x + 2)$.
2. $(4x - 3)(3x - 2)$.
3. $(6x + 1)(2x - 5)$.
4. $(9x - 2)(x + 5)$.
5. $(2x + 1)(x + 1)$.
6. $(3x - 1)(x - 1)$.
7. $(2x + 1)(x + 2)$.
8. $(3x + 1)(2x + 1)$.
9. $(2x - 3)(2x - 1)$.
10. $(5x - 1)(x - 1)$.
11. $(2x - 3)(3x - 1)$.
12. $(4x + 1)(3x + 2)$.
13. $(2a - 1)(a + 1)$.
14. $(2a + 1)(a - 1)$.
15. $(2a + 3)(a - 2)$.
16. $(5b + 2)(2b - 1)$.
17. $(5b - 2)(2b + 1)$.
18. $(2y - 5)(4y + 3)$.
19. $(4x - 1)(3x + 2)$.
20. $(7c + 2)(2c - 3)$.

p. 137. Exercise 28.

1. $(p + q)^2$.
2. $(x - 2y)^2$.
3. $(3x + 1)^2$.
4. $(4x - 5y)^2$.
5. $(x + \frac{1}{2})^2$.
6. $\left(\frac{a}{3} + \frac{b}{2}\right)^2$.
7. $(a + b + 2)^2$.
8. $(x - y - 5)^2$.
9. $(x + 10)(x - 10)$.
10. $(ab + 5)(ab - 5)$.
11. $(2x + 3y)(2x - 3y)$.
12. $(5a + 4b)(5a - 4b)$.
13. $(11x + 6y)(11x - 6y)$.
14. $(12p + 13q)(12p - 13q)$.
15. $(5 + 4a)(5 - 4a)$.
16. $(1 + 15x)(1 - 15x)$.
17. $2(2a + 5b)(2a - 5b)$.
18. $3(x + 5)(x - 5)$.
19. $5(x + 3y)(x - 3y)$.
20. $(a + b + c)(a + b - c)$.
21. $(x + 2y + 4z)(x + 2y - 4z)$.
22. $(1 + \frac{9}{4}y)(1 - \frac{9}{4}y)$.
23. $(x + y + z)(x - y - z)$.
24. $(a + x - 2y)(a - x + 2y)$.
25. $(x - 15)(x - 1)$.
26. $4ab$.
27. 3000.
28. 2100.
29. 880.
30. 252.
31. 15.
32. 150.
33. 264.
34. 68.
35. $38\cdot4$.
36. 630.
37. 140.
38. (1) 20,   (2) 60.

p. 139. Exercise 29.

1. $(x + c)(x^2 - cx + c^2)$.
2. $(y - a)(y^2 + ay + a^2)$.
3. $(1 + 2a)(1 - 2a + 4a^2)$.
4. $(x - 4)(x^2 + 4x + 16)$.
5. $(2 + 3c)(4 - 6c + 9c^2)$.
6. $(R - 1)(R^2 + R + 1)$.
7. $(m - 5n)(m^2 + 5mn + 25n^2)$.
8. $(xy + \frac{1}{2})\left(x^2y^2 - \frac{xy}{2} + \frac{1}{4}\right)$.
9. $\left(\frac{1}{x} + \frac{1}{y}\right)\left(\frac{1}{x^2} - \frac{1}{xy} + \frac{1}{y^2}\right)$.
10. $\left(\frac{1}{x} - \frac{1}{y}\right)\left(\frac{1}{x^2} + \frac{1}{xy} + \frac{1}{y^2}\right)$.

p. 141. Exercise 30.

1. $\dfrac{y}{3xz}$.
2. $\dfrac{3d^2}{4a^2}$.
3. $\dfrac{x(x + y)}{2y(x - y)}$.
4. $\dfrac{a^2 + 3b}{2ab(a - b)}$.
5. $\dfrac{3}{2xy}$.
6. $\dfrac{a + b}{a}$.
7. $\dfrac{2}{x - 2}$.
8. $\dfrac{x - 5}{x + 5}$.
9. $\dfrac{a(a + 3b)}{b(a + 2b)}$.
10. $\dfrac{x + y}{x^2 + xy + y^2}$.

p. 142. Exercise 31.

1. $\dfrac{x^2}{(x+1)}$.

2. $y(y+1)$.

3. $\dfrac{b}{a}$.

4. $\dfrac{a-7}{a-3}$.

5. $\dfrac{a(a+2b)}{a-b}$.

6. $\dfrac{(x+1)(x-3)}{x}$.

7. 1.

8. $\dfrac{1}{2(x-1)}$.

9. $\dfrac{2(b-4)}{b(b-1)}$; $\dfrac{1}{10}$.

10. $\dfrac{x+3}{2}$; $2{\cdot}25$.

p. 145. Exercise 32.

1. $\dfrac{2x-y}{x(x-y)}$.

2. $\dfrac{-(x+1)}{(x-1)(x-2)}$.

3. $\dfrac{13a-5b}{6(a^2-b^2)}$.

4. $\dfrac{x^2+2x-2}{x(x+2)}$.

5. $\dfrac{3x^2-20x+34}{(x-3)^2}$.

6. $\dfrac{-5x^2-17x}{6(x+1)(x-2)}$.

7. $\dfrac{7-3a}{(1-a)^2}$.

8. $\dfrac{12x}{x^2-9}$.

9. $\dfrac{2y-x^2-2xy}{x(x+2y)}$.

10. $\dfrac{a(10b-3a)}{3b(3a-b)}$.

11. $\dfrac{t(a-b)}{(1-at)(1-bt)}$.

12. $\dfrac{(x-a)+t(y-b)}{(a+bt)(x+yt)}$.

13. $\dfrac{2x-1}{x^2-y^2}$.

14. $\dfrac{-2y^2}{(x-y)^2(x+y)}$.

15. $R=\dfrac{pq}{p+q}$.

16. $R=\dfrac{pq}{q-p}$.

17. $R=\dfrac{R_1R_2}{2R_2+3R_1}$.

18. $R=\dfrac{R_1R_2}{3R_2-2R_1}$.

19. $R=-\dfrac{r^2-s^2}{2s}$.

20. $\dfrac{5(p^2-q^2)}{5p-q}$.

21. $\dfrac{2PQ}{P+Q}$.

22. $\dfrac{Q(P+Q)}{P-Q}$.

23. $1{\cdot}47$.

24. $P=\dfrac{2b+a}{2a+b}\cdot Q$.

25. $n=\dfrac{IR}{E-Ir}$; 6.

p. 147.   Exercise 33.

1. 3.   2. $\frac{5}{8}$.   3. 2·5.
4. $-\frac{3}{8}$.   5. $8\frac{3}{13}$.   6. $\frac{6}{8}$.
7. 7.   8. 0·3.   9. 4.
10. 6.   11. $\frac{8}{8}$.   12. 6.

13. 8.   14. $V = \dfrac{RE}{R + r}$.

p. 163.   Exercise 34.

Answers involving decimals are mostly approximate.

1. (1) 5·3.   (2) ± 2·65.   (3) ± 1·87.
2. (1) ± 1·8.   (2) ± 2·83.   (3) ± 3·46.
3. (1) ± 2·45.   (2) ± 2·24.   (3) ± 1·41.
4. (1) ± 1·41.   (2) ± 2·45.
5. (1) 3·73 or 0·27.   (2) — 4 or — 2.
   (3) 4·24 or — 0·24.
6. (1) — 3·41 or — 0·59.   (2) — 3 or — 1.
   (3) 0·83 or — 4·83.
7. Min. value is — 4 when $x = 3$.
   (1) 5 or 1.   (2) 6·16 or — 0·16.
8. Min. value is — 2 when $x = 2$.
   (1) 4·45 or — 0·45.   (2) 5 or — 1.
9. Min. value is — 1·1 approx. when $x = \frac{5}{4}$. On substituting
   $x = \frac{5}{4}$ in the function, this equals — $1\frac{1}{8}$.
   (1) 2 or $\frac{1}{2}$.   (2) 2·69 or — 0·19.
10. Max. value is 2·25 when $x = — 0·5$.

p. 169.   Exercise 35.

1. ± 2.   2. ± $\frac{1}{2}$.
3. ± 12.   4. 1, — 3.
5. 8, — 2.   6. 1, — 11.
7. $\frac{3}{4}$, — $\frac{5}{4}$.   8. 16·75, 1·25.
9. 8, 2.   10. 3, — 4.
11. 5, — 3.   12. 4, — 7.
13. 8, — 4.   14. 2, $\frac{3}{2}$.
15. $\frac{5}{2}$, — 1.   16. 3·25, — 0·92.
17. 1·434, 0·232.   18. 4 and — 1·5.
19. 5, — $\frac{1}{3}$.   20. 2·732, — 0·732.
21. 13·14, — 1·14.   22. 6·59, — 7·59.

p. 171.   Exercise 36.

1. 0, 3.   2. 0, — 5.   3. 2, 2.
4. 1, 2.   5. 1, — 4.   6. 3, — 7.
7. $\frac{7}{3}$, — $\frac{11}{2}$.   8. 4, 5.   9. 2, — 3.

10. 5, — 7. 11. — 3, — 10. 12. 11, — 6.
13. 9, — 5. 14. $\frac{3}{5}$, 4. 15. $\frac{5}{3}$, — 1.
16. 1, $\frac{1}{3}$. 17. $\frac{5}{3}$, — $\frac{3}{4}$. 18. $\frac{1}{4}$, — $\frac{2}{3}$.
19. 2, 4, 5. 20. $\frac{1}{2}$, — 3, — 2. 21. 1, 4, — 2.
22. $\frac{4}{5}$, 10, — 5. 23. a, b. 24. $\frac{c}{2}$, — d.

p. 175. Exercise 37.

Answers are approximate.

1. $0\cdot303$, — $3\cdot303$. 2. $4\cdot561$, $0\cdot438$.
3. $1\cdot48$, — $1\cdot08$. 4. $1\cdot43$, $0\cdot23$.
5. $2\cdot85$, — $0\cdot35$. 6. $1\cdot744$, — $0\cdot344$.
7. $0\cdot535$, — $6\cdot535$. 8. $3\cdot441$, — $0\cdot775$.
9. $0\cdot427$, — $2\cdot927$. 10. $0\cdot803$, — $2\cdot803$.
11. $1\cdot081$, — $1\cdot481$. 12. — $0\cdot130$, — $0\cdot770$.
13. $2\cdot591$, — $0\cdot257$. 14. $3\cdot68$, — $0\cdot43$.
15. $9\cdot75$, — $1\cdot75$. 16. $2\cdot13$, — $5\cdot62$.
17. $6\cdot123$, — $2\cdot123$.

p. 177. Exercise 38.

1. $\frac{5}{2}$ or $\frac{2}{5}$. 2. 9 mm, 15 mm.
3. $6\cdot53$ or — $1\cdot53$. 4. $21\cdot37$, $27\cdot37$.
5. $5\cdot55$. Negative root has no meaning for the problem.
6. $79\cdot78$. Second root is inadmissible.
7. $12\cdot5$, — 10.
8. 8. Negative root is inadmissible.
9. 15 mm, 20 mm. 10. 16 m², 25 m².
11. 7, — 1.
12. 12. Negative root is inadmissible.
13. 108 m². 14. $78\cdot4$ m.
15. 8 and 12, or — 12 and — 8. 16. 12.

p. 180. Exercise 39.

1. $x = 6, y = 4$; $x = — 5, y = — 7$.
2. $x = 5, y = 2$; $x = — 26, y = 33$.
3. $x = 2, y = 1$; $x = — \frac{7}{11}, y = \frac{90}{11}$.
4. $x = \frac{1}{2}, y = 6\frac{1}{2}$; $x = 3\frac{1}{2}, y = — 2\frac{1}{2}$.
5. $x = — 3, y = 2$; $x = — 2, y = 1$.
6. $x = 1, y = 4$; $x = \frac{4}{3}, y = \frac{34}{9}$.
7. $x = 4, y = 3$; $x = — \frac{15}{14}, y = — \frac{171}{14}$.
8. $x = — 14, y = 11$; $x = \frac{8}{3}, y = — \frac{1}{3}$.
9. $x = 7, y = 4$; $x = \frac{1}{3}, y = 21$.
10. $x = 6, y = 4$; $x = — \frac{6}{13}, y = — \frac{110}{13}$.
11. $x = 2, y = 4$; $x = \frac{4}{3}, y = — \frac{2}{3}$.

p. 182.   **Exercise 40.**

1. $x = 3, y = 5$; $x = 5, y = 3$.
2. $x = 7, y = 2$; $x = 2, y = 7$.
3. $x = 8, y = 3$; $x = -3, y = -8$.
4. $x = \pm 4, y = \pm 1$; $x = \pm 1, y = \pm 4$.
5. $x = 6, y = 2$; $x = -4, y = -3$.
6. $x = 4, y = 1$; $x = -1, y = -4$.
7. $x = 9, y = 2$; $x = 2, \; y = 9$; $x = -2, \; y = -9$;
       $x = -9, y = -2$.
8. $x = 7, y = 1$; $x = 1, y = 7$.
9. $x = 6, \; y = 3$; $x = -6, \; y = -3$; $x = 3, \; y = 6$;
       $x = -3, y = -6$.
10. $x = 3, y = 1$; $x = 1{\cdot}5, y = 2$.

p. 186.   **Exercise 41.**

1. $6a^9$.
2. a^{12}.
3. $\frac{1}{8}a^8$.
4. $2^{10} = 1024$.
5. $4^6 = 4096$.
6. $6a^8b^7$.
7. $x^6y^{12}z^4$.
8. x^{2m}.
9. a^{2m}.
10. $a^{2p+q}b^p$.
11. a^{2n}.
12. a^{m+n+4}.
13. a^3.
14. $\frac{1}{2}x^3$.
15. $3a^5$.
16. $2^6 = 64$.
17. $3^3 = 27$.
18. x^5.
19. $-5a^3b$.
20. a^p.
21. $4x^py^q$.
22. a^7.
23. x.
24. a^{n+1}.
25. a^{2n}.
26. a^6.
27. $\dfrac{8ab}{9}$.
28. $\dfrac{a^2}{b^2}$.
29. $2^6 = 64$.
30. $3^6 = 729$.
31. x^{10}.
32. x^{10}.
33. a^{16}.
34. $4a^4b^6$.
35. $\dfrac{x^{12}y^9}{8}$.
36. a^{3p}.
37. x^{4n}.
38. $27a^{6p}$.
39. a^2.
40. x^4.
41. x^8.
42. $3a^3$.
43. a^2b.
44. $\dfrac{x^3}{y^2}$.
45. $\dfrac{3a^4}{2b^3}$.
46. x^n.

p. 193.   **Exercise 42.**

1. (1) $\sqrt[4]{3}$.
   (2) $\sqrt[5]{8}$.
   (3) $\sqrt[3]{a}$.
   (4) $\sqrt{2}$.
   (5) $\sqrt[4]{3}$.
   (6) $\sqrt[4]{a^3}$.
2. (1) $\sqrt[4]{1000}$.
   (2) $\sqrt[3]{16}$.
   (3) $\sqrt[4]{a^5}$.
   (4) $10\sqrt{10}$.
   (5) $4\sqrt{2}$.
   (6) $10\sqrt[4]{10}$.
3. (1) $2\sqrt[5]{2}$.
   (2) $4\sqrt[4]{2}$.
   (3) $\sqrt[4]{a^7}$.
   (4) $\sqrt{a^7} = a^3\sqrt{a}$.
   (5) $\sqrt[5]{a^6} = a\sqrt[5]{a}$.
   (6) $\sqrt[5]{a}$.

4. (1) 8.   (2) 32.   (3) 27.
  (4) 1000.   (5) 31·6.   (6) $\frac{1}{2}$.

5. $\frac{1}{9}$, $\frac{1}{3}$, 1, 3, $3\sqrt{3}$, 9, $9\sqrt{3}$.

6. (1) 5·656.   (2) 27.   (3) $\frac{1}{10}$.
  (4) $a^{1\frac{1}{2}}$.   (5) 2.   (6) $a^{1\cdot2}$

7. $\frac{1}{4}$, $\dfrac{3}{a^2}$, $\dfrac{1}{\sqrt{2}}$, 3, $3a^2$, $\dfrac{1}{1000}$.

8. (1) $\dfrac{1}{x^3}$.   (2) $\dfrac{1}{x^{\frac{1}{2}}}$.   (3) x^3.

  (4) $\dfrac{1}{x^6}$.   (5) x.   (6) $\frac{1}{2}x^{\frac{1}{2}}$.

9. (1) 4.   (2) 125.   (3) 1000.

  (4) $\dfrac{1}{5^6} = \dfrac{1}{156\ 25}$.   (5) 16.   (6) 3.

10. (1) 4.   (2) $\frac{27}{8}$.   (3) $\frac{1}{4}$.
   (4) $\frac{1}{8}$.   (5) 8.   (6) $\frac{1}{32}$.

11. 5·656.

12. (1) $\sqrt[6]{a^5}$.   (2) $100\sqrt{10} = 316\cdot2$.   (3) $\dfrac{1}{a^3}$.

   (4) $\dfrac{1}{a^2}$.   (5) x^{3n}.   (6) x^{2n}.

13. (1) a^3.   (2) 1.   (3) 0.
14. (1) $3\cdot52 \times 10^9$.   (2) $1\cdot633 \times 10^7$.   (3) $2\cdot2 \times 10^2$.
   (4) $3\cdot7 \times 10^3$.   (5) $2\cdot52 \times 10^{-7}$.

p. 202.   Exercise 43.

1. 1, 3, 4, 2, 0, 5, 1, 3, 0, 2.
2. (1) 0·6990, 1·6990, 2·6990, 4·6990.
  (2) 0·6721, 2·6721, 4·6721.
  (3) 1·7226, 0·7226, 2·7226.
  (4) 2·9767, 0·9767, 4·9767.
  (5) 0·7588, 1·9842, 3·8433.
3. (1) 446·7, 446 70, 44·67.   (2) 87·70, 8770, 8·770.
  (3) 4·714, 471·4, 471 400.   (4) 2628, 5·229, 114·0.

p. 204.   Exercise 44.

1. 344·6.   2. 276·4.   3. 1397.
4. 5975.   5. 2·396.   6. 6·997.
7. 1·589.   8. 222·8.   9. 14·22.
10. 13·56.   11. 851·3.   12. 2650.
13. 3·137.   14. 728·5.   15. 2·172.
16. 104·6.   17. 1·656.   18. 1436.
19. 1·359.   20. 1·695.   21. 2·321.
22. 2·786.   23. 5·002.   24. 1·546.

p. 206.   **Exercise 45.**

1. (1) $0\cdot4470$, $\bar{1}\cdot4470$, $\bar{2}\cdot4470$.   (2) $0\cdot6298$, $\bar{1}\cdot6298$, $\bar{3}\cdot6298$.
   (3) $\bar{3}\cdot9904$, $\bar{4}\cdot9904$, $\bar{1}\cdot9904$.   (4) $\bar{2}\cdot8097$, $\bar{1}\cdot8087$, $\bar{4}\cdot8097$.

2. (1) $\bar{2}\cdot7771$.   (2) $\bar{4}\cdot6749$.   (3) $\bar{3}\cdot9011$.
   (4) $\bar{5}\cdot9673$.   (5) $\bar{1}\cdot7538$.   (6) $\bar{2}\cdot9023$.

3. (1) $0\cdot2159$.   (2) $0\cdot007\ 453$.   (3) $0\cdot030\ 70$.
   (4) $0\cdot000\ 440\ 2$.   (5) $0\cdot5940$.   (6) $2\cdot482 \times 10^{-8}$.

p. 209.   **Exercise 46.**

1. (1) $\bar{4}\cdot6037$.   (2) $\bar{2}\cdot7126$.
2. (1) $\bar{2}\cdot5926$.   (2) $0\cdot8263$.   (3) $\bar{1}\cdot6597$.
   (4) $2\cdot4814$.
3. (1) $\bar{1}\cdot7464$.   (2) $\bar{4}\cdot8368$.   (3) $\bar{3}\cdot8910$.
   (4) $\bar{1}\cdot3673$.   (5) $\bar{1}\cdot1958$.   (6) $\bar{4}\cdot7913$.
4. (1) $\bar{2}\cdot6856$.   (2) $\bar{1}\cdot071\ 55$.   (3) $\bar{1}\cdot7754$.
   (4) $\bar{1}\cdot1463$.   (5) $\bar{2}\cdot0254$.   (6) $0\cdot5619$.
5. (1) $\bar{1}\cdot7399$.   (2) $\bar{1}\cdot7127$.   (3) $\bar{2}\cdot7726$.
   (4) $\bar{2}\cdot5598$.   (5) $\bar{1}\cdot7266$.   (6) $\bar{3}\cdot8973$.

p. 211.   **Exercise 47.**

1. $15\cdot42$.   2. $0\cdot3285$.   3. $0\cdot015\ 29$.
4. $5\cdot699$.   5. $0\cdot6116$.   6. $0\cdot032\ 39$.
7. $0\cdot049\ 03$.   8. $0\cdot1600$.   9. $85\cdot23$.
10. $0\cdot8414$.   11. $0\cdot1226$.   12. $1\cdot197$.
13. $0\cdot071\ 15$.   14. $1\cdot826$.   15. $1\cdot457$.
16. $3\cdot558$.   17. $5\cdot471$.   18. $0\cdot1014$.
19. $0\cdot1429$.   20. $9\cdot399$.   21. $0\cdot3609$.
22. $0\cdot031\ 29$.   23. $1\cdot022$.   24. $62\cdot84$.
25. $1\cdot663$.   26. $3\cdot940$.   27. $9\cdot527$.
28. $896\cdot6$.   29. $102\cdot2$.   30. $516\cdot4$.
31. $2\cdot89$.   32. $1\cdot035$.   33. $3\cdot98$.
34. $0\cdot795$.   35. $9\cdot81$.   36. (a) 3; (b) 3.
37. $530\cdot7$.   38. $4\cdot923$ mm.   39. (1) $1\cdot5261$,
40. 1200.   41. $0\cdot8268$.   (2) $\bar{3}\cdot2193$.

p. 218.   **Exercise 48.**

1. (a) $100a : b$.   (b) $\dfrac{3600p}{60q + r}$.   2. $36 \times 10^5 : 1$.

3. (1) $\frac{8}{5}$.   (2) $25 : 64$.   (3) $5 : 12$.

4. (a) 35.   (b) $\dfrac{bx}{a}$.   (c) $\dfrac{a}{x}$.

5. $\dfrac{c(b + c)}{a}$.   6. $\dfrac{pA}{p + q}$, $\dfrac{qA}{p + q}$.

7. $\dfrac{ax}{x + y}$ kg, $\dfrac{ay}{x + y}$ kg.   8. $\dfrac{bc}{a}$.   9. 17.

10. (1) $b\sqrt{ac}$.   (2) $4ab$.   (3) $(a + b)\sqrt{ab}$.

11. 7. 12. $\frac{7}{8}$. 13. (1) $\frac{7}{8}$. (2) $\frac{8}{9}$. (3) $\frac{5}{4}$.

14. (a) 450. (b) $\dfrac{6a}{5b}$. (c) $\dfrac{4a}{3b}$.

15. $x = \sqrt[3]{a^2b},\ y = \sqrt[3]{ab^2}$.

p. 226. Exercise 49.

1. (a) No. The runner's rate differs at various parts of the race.
   (b) Yes. (c) No. (d) Yes.
   (e) No. The connecting law is as explained in § 204.

2. $\frac{8}{7}$; $45\frac{5}{7}$. 3. 37·5; 3·36.

4. $y = \frac{33}{7}x$; $4\frac{5}{11}$. 5. $s = 3·2t$; 8·96 m.

6. $E = \frac{8}{15}W$; 3·73. 7. $y = \frac{1}{2}x + 6$.

8. $E = 0·06W + 0·2$. 9. $F = 0·42W - 0·6$.

p. 234. Exercise 50.

1. $y = \frac{8}{9}x^2$; $64\frac{2}{9}$. 2. $y = \frac{1}{4}x^3$; $\frac{27}{4}$.

3. y values 12, 3; x value 6. 4. $y = \frac{7}{4}\sqrt{x}$; 8·75.

5. 384; $2\sqrt[3]{4}$. 6. $y = \dfrac{30}{x}$; 1·5.

7. x values 2, 15; y value 10. 8. $F = \dfrac{1920}{d^2}$.

9. 4 s. 10. 10 m.

11. 102 mm. 12. 1·8.

13. 50 s. 14. $y = 8x^{1·4}$.

p. 239. Exercise 51.

1. (a) $y \propto xz$. ∴ $y = kxz$.

   (b) $y \propto \dfrac{x}{z^2}$. ∴ $y = \dfrac{kx}{z^2}$.

   (c) $y \propto \dfrac{\sqrt{x}}{z}$. ∴ $y = \dfrac{k\sqrt{x}}{z}$.

   (d) $V \propto hr^2$. ∴ $y = khr^2$.

   (e) $W \propto \dfrac{bd^2}{l}$. ∴ $W = k \cdot \dfrac{bd^2}{l}$.

   (f) $y \propto \dfrac{x^2}{3\sqrt{z}}$. ∴ $y = k \cdot \dfrac{x^2}{3\sqrt{z}}$.

2. $y = \dfrac{25}{4}\dfrac{x}{z}$; 15. 3. $y = 3xz^2$; 8.

4. $10\frac{5}{12}$. 5. 8 t. 6. 1.

7. $H = \dfrac{0·24tE^2}{R}$. (1) 24 000 units. (2) 150 s.

8. 4·1% increase.

p. 246.   Exercise 52.

1. $y = 6x^2 + 3$.   2. $y = 0.42x^2 + 7.5$.
3. $R = 0.01V^2 + 5$.   4. $H = 1.08V^3 + 420$.

5. $y = 0.54 + 0.24x^2$.   6. $R = \dfrac{1200}{V} + 50$.

7. $y = 0.25x^3$.   8. $y = 3x^{\frac{1}{2}}$.
9. $Q = 5.7H^{0.36}$.   10. $v = 3.65x^{\frac{1}{2}}$.

p. 252.   Exercise 53.

1. (1) $\sqrt{150}$.   (2) $\sqrt{288}$.   (3) $\sqrt{500}$.
   (4) $\sqrt{208}$.   (5) $\sqrt{4a^2b}$.   (6) $\sqrt{9x^4y}$.
2. (1) $20\sqrt{2}$.   (2) $8\sqrt{5}$.   (3) $6\sqrt{3}$.
   (4) $50\sqrt{2}$.   (5) $5\sqrt{15}$.   (6) $60\sqrt{2}$.
   (7) $2ab\sqrt{6a}$.   (8) $3x^2y\sqrt{2yz}$.   (9) $5abc\sqrt{3abc}$.
   (10) $10p\sqrt{10q}$.   (11) $a(a-b)\sqrt{b(a-b)}$.
   (12) $5y(x-2y)\sqrt{xy}$.
3. (1) $3\sqrt{5}$.   (2) $7\sqrt{2}$.   (3) $16\sqrt{3}$.
   (4) $10\sqrt{10}$.   (5) $4-\sqrt{2}$.   (6) $7\sqrt{2}+\sqrt{14}$.
   (7) $(a+b)\sqrt{a-b}$.   (8) $3(x-2y)\sqrt{2(x+2y)}$.
4. (1) $2\sqrt{2}-1$.   (2) $2-3\sqrt{6}$.   (3) $3-2\sqrt{2}$.
   (4) $23+4\sqrt{15}$.   (5) $57-10\sqrt{14}$.   (6) 2.
   (7) 38.   (8) $301+20\sqrt{3}$.   (9) $a-25$.
   (10) 21.

5. (1) $\dfrac{2\sqrt{3}}{3}$.   (2) $\dfrac{\sqrt{5}}{5}$.   (3) $\dfrac{3\sqrt{7}}{14}$.

   (4) $\dfrac{\sqrt{10}}{30}$.   (5) $\dfrac{\sqrt{2}}{8}$.   (6) $\dfrac{6\sqrt{5}}{5}$.

   (7) $\dfrac{\sqrt{6}}{3}$.   (8) $\dfrac{2\sqrt{35}}{7}$.   (9) $\dfrac{\sqrt{5}+1}{4}$.

   (10) $\dfrac{3(\sqrt{7}-\sqrt{2})}{5}$.   (11) $\dfrac{3(2\sqrt{3}+\sqrt{2})}{10}$.

   (12) $5(2\sqrt{7}+3\sqrt{3})$.   (13) $5+2\sqrt{6}$.

   (14) $6-\sqrt{35}$.   (15) $\dfrac{21+\sqrt{7}}{62}$.

   (16) $2\sqrt{2}$.   (17) $-\sqrt{3}$.

   (18) $\dfrac{4(6+\sqrt{5})}{31}$.

p. 259.   Exercise 54.

1. (1) 12.5, 15, 17.5.   (2) 0, -4, -8.
   (3) $(a-3b)$, $(a-5b)$, $(a-7b)$.
   (4) 6.6, 7.9, 9.2.
   (5) $(x+2y)$, $(x+3y)$, $(x+4y)$.

2. 12, 16·5.   3. $4 + 4p$.
4. $x + 3n - 1$.   5. 3·48.
6. 29.   7. 2.   8. 19.
9. (a) 217·5.   (b) 16.   (c) 0.   (d) $364\frac{1}{2}$.
10. 12.   11. 5 or 8.   12. 120; 1860.
13. £40·05; £5343·75.   14. £34.

p. 263.   Exercise 55.

1. (a) 62·5, 156·25, 390·625.   (b) $\frac{1}{4}$, $\frac{1}{16}$, $\frac{1}{64}$.
   (c) $- 54, 81, - \frac{243}{2}$.
   (d) 3×10^{-4}, 3×10^{-5}, 3×10^{-6}.
   (e) 0·010 125, 0·001 518 75, 0·000 227 8 (approx.).
2. 320.   3. $\frac{128}{234}$.
4. 1·610 51.   5. 0·001 215.
6. (a) ar^{2n-1}, ar^{2n}; $- ar^{2n-1}$, ar^{2n}.
7. 1·05.   8. 1·04.
9. (1) 3·873; (2) 3·888 (both approx.).
10. 7, 9·8.   11. £1464·10.   12. £162 901·25.
13. 1·04.   14. 10, 20, 40.

p. 266.   Exercise 56.

1. (a) 94·5.   (b) 19·93.   (c) $\frac{21}{64}$.
2. $\frac{315}{32}$.   3. $\frac{133}{243}$.   4. 16·32.
5. 216·8.   6. 765.   7. 27; $40\frac{40}{81}$.
8. 93·8 approx.

p. 271.   Exercise 57.

1. (a) $\frac{2}{3}$.   (b) $\frac{1}{3}$.   (c) 1.
2. (a) $\frac{16}{99}$.   (b) $\frac{n}{99}$.   (c) $\frac{16}{99}$.
   Series (c) is the sum of (a) and (b). It represents
   0·161 616 . . ., a recurring decimal.
3. $\frac{2}{3}$. This series equals the series (b) in question (1) increased
   by unity and then diminished by series (a).
4. In $\frac{2}{3}\{1 - (\frac{1}{3})^n\}$, $(\frac{1}{3})^n \longrightarrow o$ as $n \longrightarrow \infty$. ∴ the limit is $\frac{2}{3}$.
5. $\dfrac{b}{b - a}\left\{1 - \left(\dfrac{a}{b}\right)^n\right\}$; limit $= \dfrac{b}{b - a}$.
6. (a) $\frac{16}{5}$.   (b) $\frac{25}{6}$.   (c) $\frac{27}{6}$.
7. (a) 26.   (b) $2(\sqrt{2} + 1)$.
8. $\frac{4}{5}$.   9. 180 m.   10. £75 000.

p. 276.   Exercise 58.

Most of the answers are approximate.

1. £1180.   2. £5500.   3. £486.
4. £2430.   5. £3431.   6. £5000.
7. £1667.   8. £240 000.

LOGARITHMS of numbers 100 to 549

	0	1	2	3	4	5	6	7	8	9	1	2	3	4	5	6	7	8	9
10	0000	0043	0086	0128	0170	0212	0253	0294	0334	0374	4	8	12	17	21	25	29	33	37
11	0414	0453	0492	0531	0569	0607	0645	0682	0719	0755	4	8	11	15	19	23	26	30	34
12	0792	0828	0864	0899	0934	0969	1004	1038	1072	1106	3	7	10	14	17	21	24	28	31
13	1139	1173	1206	1239	1271	1303	1335	1367	1399	1430	3	6	10	13	16	19	23	26	29
14	1461	1492	1523	1553	1584	1614	1644	1673	1703	1732	3	6	9	12	15	18	21	24	27
15	1761	1790	1818	1847	1875	1903	1931	1959	1987	2014	3	6	8	11	14	17	20	22	25
16	2041	2068	2095	2122	2148	2175	2201	2227	2253	2279	3	5	8	11	13	16	18	21	24
17	2304	2330	2355	2380	2405	2430	2455	2480	2504	2529	2	5	7	10	12	15	17	20	22
18	2553	2577	2601	2625	2648	2672	2695	2718	2742	2765	2	5	7	9	12	14	16	19	21
19	2788	2810	2833	2856	2878	2900	2923	2945	2967	2989	2	4	7	9	11	13	16	18	20
20	3010	3032	3054	3075	3096	3118	3139	3160	3181	3201	2	4	6	8	11	13	15	17	19
21	3222	3243	3263	3284	3304	3324	3345	3365	3385	3404	2	4	6	8	10	12	14	16	18
22	3424	3444	3464	3483	3502	3522	3541	3560	3579	3598	2	4	6	8	10	12	14	15	17
23	3617	3636	3655	3674	3692	3711	3729	3747	3766	3784	2	4	6	7	9	11	13	15	17
24	3802	3820	3838	3856	3874	3892	3909	3927	3945	3962	2	4	5	7	9	11	12	14	16
25	3979	3997	4014	4031	4048	4065	4082	4099	4116	4133	2	3	5	7	9	10	12	14	15
26	4150	4166	4183	4200	4216	4232	4249	4265	4281	4298	2	3	5	7	8	10	11	13	15
27	4314	4330	4346	4362	4378	4393	4409	4425	4440	4456	2	3	5	6	8	9	11	13	14
28	4472	4487	4502	4518	4533	4548	4564	4579	4594	4609	2	3	5	6	8	9	11	12	14
29	4624	4639	4654	4669	4683	4698	4713	4728	4742	4757	1	3	4	6	7	9	10	12	13
30	4771	4786	4800	4814	4829	4843	4857	4871	4886	4900	1	3	4	6	7	9	10	11	13
31	4914	4928	4942	4955	4969	4983	4997	5011	5024	5038	1	3	4	5	7	8	10	11	12
32	5051	5065	5079	5092	5105	5119	5132	5145	5159	5172	1	3	4	5	7	8	9	11	12
33	5185	5198	5211	5224	5237	5250	5263	5276	5289	5302	1	3	4	5	6	8	9	10	12
34	5315	5328	5340	5353	5366	5378	5391	5403	5416	5428	1	3	4	5	6	8	9	10	11
35	5441	5453	5465	5478	5490	5502	5514	5527	5539	5551	1	2	4	5	6	7	9	10	11
36	5563	5575	5587	5599	5611	5623	5635	5647	5658	5670	1	2	4	5	6	7	8	10	11
37	5682	5694	5705	5717	5729	5740	5752	5763	5775	5786	1	2	3	5	6	7	8	9	10
38	5798	5809	5821	5832	5843	5855	5866	5877	5888	5899	1	2	3	5	6	7	8	9	10
39	5911	5922	5933	5944	5955	5966	5977	5988	5999	6010	1	2	3	4	5	7	8	9	10
40	6021	6031	6042	6053	6064	6075	6085	6096	6107	6117	1	2	3	4	5	6	7	9	10
41	6128	6138	6149	6160	6170	6180	6191	6201	6212	6222	1	2	3	4	5	6	7	8	9
42	6232	6243	6253	6263	6274	6284	6294	6304	6314	6325	1	2	3	4	5	6	7	8	9
43	6335	6345	6355	6365	6375	6385	6395	6405	6415	6425	1	2	3	4	5	6	7	8	9
44	6435	6444	6454	6464	6474	6484	6493	6503	6513	6522	1	2	3	4	5	6	7	8	9
45	6532	6542	6551	6561	6571	6580	6590	6599	6609	6618	1	2	3	4	5	6	7	8	9
46	6628	6637	6646	6656	6665	6675	6684	6693	6702	6712	1	2	3	4	5	6	7	7	8
47	6721	6730	6739	6749	6758	6767	6776	6785	6794	6803	1	2	3	4	5	5	6	7	8
48	6812	6821	6830	6839	6848	6857	6866	6875	6884	6893	1	2	3	4	4	5	6	7	8
49	6902	6911	6920	6928	6937	6946	6955	6964	6972	6981	1	2	3	4	4	5	6	7	8
50	6990	6998	7007	7016	7024	7033	7042	7050	7059	7067	1	2	3	3	4	5	6	7	8
51	7076	7084	7093	7101	7110	7118	7126	7135	7143	7152	1	2	3	3	4	5	6	7	8
52	7160	7168	7177	7185	7193	7202	7210	7218	7226	7235	1	2	2	3	4	5	6	7	7
53	7243	7251	7259	7267	7275	7284	7292	7300	7308	7316	1	2	2	3	4	5	6	6	7
54	7324	7332	7340	7348	7356	7364	7372	7380	7388	7396	1	2	2	3	4	5	6	6	7
	0	1	2	3	4	5	6	7	8	9	1	2	3	4	5	6	7	8	9

LOGARITHMS of numbers 550 to 999

Proportional Parts

	0	1	2	3	4	5	6	7	8	9	1	2	3	4	5	6	7	8	9
55	7404	7412	7419	7427	7435	7443	7451	7459	7466	7474	1	2	2	3	4	5	5	6	7
56	7482	7490	7497	7505	7513	7520	7528	7536	7543	7551	1	2	2	3	4	5	5	6	7
57	7559	7566	7574	7582	7589	7597	7604	7612	7619	7627	1	2	2	3	4	5	5	6	7
58	7634	7642	7649	7657	7664	7672	7679	7686	7694	7701	1	1	2	3	4	4	5	6	7
59	7709	7716	7723	7731	7738	7745	7752	7760	7767	7774	1	1	2	3	4	4	5	6	7
60	7782	7789	7796	7803	7810	7818	7825	7832	7839	7846	1	1	2	3	4	4	5	6	6
61	7853	7860	7868	7875	7882	7889	7896	7903	7910	7917	1	1	2	3	4	4	5	6	6
62	7924	7931	7938	7945	7952	7959	7966	7973	7980	7987	1	1	2	3	4	4	5	6	6
63	7993	8000	8007	8014	8021	8028	8035	8041	8048	8055	1	1	2	3	3	4	5	6	6
64	8062	8069	8075	8082	8089	8096	8102	8109	8116	8122	1	1	2	3	3	4	5	5	6
65	8129	8136	8142	8149	8156	8162	8169	8176	8182	8189	1	1	2	3	3	4	5	5	6
66	8195	8202	8209	8215	8222	8228	8235	8241	8248	8254	1	1	2	3	3	4	5	5	6
67	8261	8267	8274	8280	8287	8293	8299	8306	8312	8319	1	1	2	3	3	4	4	5	6
68	8325	8331	8338	8344	8351	8357	8363	8370	8376	8382	1	1	2	3	3	4	4	5	6
69	8388	8395	8401	8407	8414	8420	8426	8432	8439	8445	1	1	2	3	3	4	4	5	6
70	8451	8457	8463	8470	8476	8482	8488	8494	8500	8506	1	1	2	2	3	4	4	5	6
71	8513	8519	8525	8531	8537	8543	8549	8555	8561	8567	1	1	2	2	3	4	4	5	5
72	8573	8579	8585	8591	8597	8603	8609	8615	8621	8627	1	1	2	2	3	4	4	5	5
73	8633	8639	8645	8651	8657	8663	8669	8675	8681	8686	1	1	2	2	3	4	4	5	5
74	8692	8698	8704	8710	8716	8722	8727	8733	8739	8745	1	1	2	2	3	4	4	5	5
75	8751	8756	8762	8768	8774	8779	8785	8791	8797	8802	1	1	2	2	3	3	4	5	5
76	8808	8814	8820	8825	8831	8837	8842	8848	8854	8859	1	1	2	2	3	3	4	5	5
77	8865	8871	8876	8882	8887	8893	8899	8904	8910	8915	1	1	2	2	3	3	4	4	5
78	8921	8927	8932	8938	8943	8949	8954	8960	8965	8971	1	1	2	2	3	3	4	4	5
79	8976	8982	8987	8993	8998	9004	9009	9015	9020	9025	1	1	2	2	3	3	4	4	5
80	9031	9036	9042	9047	9053	9058	9063	9069	9074	9079	1	1	2	2	3	3	4	4	5
81	9085	9090	9096	9101	9106	9112	9117	9122	9128	9133	1	1	2	2	3	3	4	4	5
82	9138	9143	9149	9154	9159	9165	9170	9175	9180	9186	1	1	2	2	3	3	4	4	5
83	9191	9196	9201	9206	9212	9217	9222	9227	9232	9238	1	1	2	2	3	3	4	4	5
84	9243	9248	9253	9258	9263	9269	9274	9279	9284	9289	1	1	2	2	3	3	4	4	5
85	9294	9299	9304	9309	9315	9320	9325	9330	9335	9340	1	1	2	2	3	3	4	4	5
86	9345	9350	9355	9360	9365	9370	9375	9380	9385	9390	1	1	2	2	3	3	4	4	5
87	9395	9400	9405	9410	9415	9420	9425	9430	9435	9440	0	1	1	2	2	3	3	4	4
88	9445	9450	9455	9460	9465	9469	9474	9479	9484	9489	0	1	1	2	2	3	3	4	4
89	9494	9499	9504	9509	9513	9518	9523	9528	9533	9538	0	1	1	2	2	3	3	4	4
90	9542	9547	9552	9557	9562	9566	9571	9576	9581	9586	0	1	1	2	2	3	3	4	4
91	9590	9595	9600	9605	9609	9614	9619	9624	9628	9633	0	1	1	2	2	3	3	4	4
92	9638	9643	9647	9652	9657	9661	9666	9671	9675	9680	0	1	1	2	2	3	3	4	4
93	9685	9689	9694	9699	9703	9708	9713	9717	9722	9727	0	1	1	2	2	3	3	4	4
94	9731	9736	9741	9745	9750	9754	9759	9764	9768	9773	0	1	1	2	2	3	3	4	4
95	9777	9782	9786	9791	9795	9800	9805	9809	9814	9818	0	1	1	2	2	3	3	4	4
96	9823	9827	9832	9836	9841	9845	9850	9854	9859	9863	0	1	1	2	2	3	3	4	4
97	9868	9872	9877	9881	9886	9890	9894	9899	9903	9908	0	1	1	2	2	3	3	4	4
98	9912	9917	9921	9926	9930	9934	9939	9943	9948	9952	0	1	1	2	2	3	3	4	4
99	9956	9961	9965	9969	9974	9978	9983	9987	9991	9996	0	1	1	2	2	3	3	4	4
	0	1	2	3	4	5	6	7	8	9	1	2	3	4	5	6	7	8	9

ANTI-LOGARITHMS

Proportional Parts

	0	1	2	3	4	5	6	7	8	9	1	2	3	4	5	6	7	8	9
·00	1000	1002	1005	1007	1009	1012	1014	1016	1019	1021	0	0	1	1	1	1	2	2	2
·01	1023	1026	1028	1030	1033	1035	1038	1040	1042	1045	0	0	1	1	1	1	2	2	2
·02	1047	1050	1052	1054	1057	1059	1062	1064	1067	1069	0	0	1	1	1	1	2	2	2
·03	1072	1074	1076	1079	1081	1084	1086	1089	1091	1094	0	0	1	1	1	1	2	2	2
·04	1096	1099	1102	1104	1107	1109	1112	1114	1117	1119	0	1	1	1	1	2	2	2	2
·05	1122	1125	1127	1130	1132	1135	1138	1140	1143	1146	0	1	1	1	1	2	2	2	2
·06	1148	1151	1153	1156	1159	1161	1164	1167	1169	1172	0	1	1	1	1	2	2	2	2
·07	1175	1178	1180	1183	1186	1189	1191	1194	1197	1199	0	1	1	1	1	2	2	2	2
·08	1202	1205	1208	1211	1213	1216	1219	1222	1225	1227	0	1	1	1	1	2	2	2	3
·09	1230	1233	1236	1239	1242	1245	1247	1250	1253	1256	0	1	1	1	1	2	2	2	3
·10	1259	1262	1265	1268	1271	1274	1276	1279	1282	1285	0	1	1	1	1	2	2	2	3
·11	1288	1291	1294	1297	1300	1303	1306	1309	1312	1315	0	1	1	1	2	2	2	2	3
·12	1318	1321	1324	1327	1330	1334	1337	1340	1343	1346	0	1	1	1	2	2	2	3	3
·13	1349	1352	1355	1358	1361	1365	1368	1371	1374	1377	0	1	1	1	2	2	2	3	3
·14	1380	1384	1387	1390	1393	1396	1400	1403	1406	1409	0	1	1	1	2	2	3	3	3
·15	1413	1416	1419	1422	1426	1429	1432	1435	1439	1442	0	1	1	1	2	2	2	3	3
·16	1445	1449	1452	1455	1459	1462	1466	1469	1472	1476	0	1	1	1	2	2	3	3	3
·17	1479	1483	1486	1489	1493	1496	1500	1503	1507	1510	0	1	1	1	2	2	2	3	3
·18	1514	1517	1521	1524	1528	1531	1535	1538	1542	1545	0	1	1	1	2	2	2	3	3
·19	1549	1552	1556	1560	1563	1567	1570	1574	1578	1581	0	1	1	1	2	2	3	3	3
·20	1585	1589	1592	1596	1600	1603	1607	1611	1614	1618	0	1	1	1	2	2	3	3	3
·21	1622	1626	1629	1633	1637	1641	1644	1648	1652	1656	0	1	1	2	2	2	3	3	3
·22	1660	1663	1667	1671	1675	1679	1683	1687	1690	1694	0	1	1	2	2	2	3	3	3
·23	1698	1702	1706	1710	1714	1718	1722	1726	1730	1734	0	1	1	2	2	2	3	3	4
·24	1738	1742	1746	1750	1754	1758	1762	1766	1770	1774	0	1	1	2	2	2	3	3	4
·25	1778	1782	1786	1791	1795	1799	1803	1807	1811	1816	0	1	1	2	2	3	3	3	4
·26	1820	1824	1828	1832	1837	1841	1845	1849	1854	1858	0	1	1	2	2	3	3	3	4
·27	1862	1866	1871	1875	1879	1884	1888	1892	1897	1901	0	1	1	2	2	3	3	3	4
·28	1905	1910	1914	1919	1923	1928	1932	1936	1941	1945	0	1	1	2	2	3	3	4	4
·29	1950	1954	1959	1963	1968	1972	1977	1982	1986	1991	0	1	1	2	2	3	3	4	4
·30	1995	2000	2004	2009	2014	2018	2023	2028	2032	2037	0	1	1	2	2	3	3	4	4
·31	2042	2046	2051	2056	2061	2065	2070	2075	2080	2084	0	1	1	2	2	3	3	4	4
·32	2089	2094	2099	2104	2109	2113	2118	2123	2128	2133	0	1	1	2	2	3	3	4	4
·33	2138	2143	2148	2153	2158	2163	2168	2173	2178	2183	0	1	1	2	2	3	3	4	4
·34	2188	2193	2198	2203	2208	2213	2218	2223	2228	2234	1	1	2	2	3	3	4	4	5
·35	2239	2244	2249	2254	2259	2265	2270	2275	2280	2286	1	1	2	2	3	3	4	4	5
·36	2291	2296	2301	2307	2312	2317	2323	2328	2333	2339	1	1	2	2	3	3	4	4	5
·37	2344	2350	2355	2360	2366	2371	2377	2382	2388	2393	1	1	2	2	3	3	4	4	5
·38	2399	2404	2410	2415	2421	2427	2432	2438	2443	2449	1	1	2	2	3	3	4	4	5
·39	2455	2460	2466	2472	2477	2483	2489	2495	2500	2506	1	1	2	2	3	3	4	5	5
·40	2512	2518	2523	2529	2535	2541	2547	2553	2559	2564	1	1	2	2	3	3	4	5	5
·41	2570	2576	2582	2588	2594	2600	2606	2612	2618	2624	1	1	2	2	3	4	4	5	5
·42	2630	2636	2642	2648	2655	2661	2667	2673	2679	2685	1	1	2	2	3	4	4	5	6
·43	2692	2698	2704	2710	2716	2723	2729	2735	2742	2748	1	1	2	2	3	4	4	5	6
·44	2754	2761	2767	2773	2780	2786	2793	2799	2805	2812	1	1	2	3	3	4	4	5	6
·45	2818	2825	2831	2838	2844	2851	2858	2864	2871	2877	1	1	2	3	3	4	5	5	6
·46	2884	2891	2897	2904	2911	2917	2924	2931	2938	2944	1	1	2	3	3	4	5	5	6
·47	2951	2958	2965	2972	2979	2985	2992	2999	3006	3013	1	1	2	3	3	4	5	6	6
·48	3020	3027	3034	3041	3048	3055	3062	3069	3076	3083	1	1	2	3	4	4	5	6	6
·49	3090	3097	3105	3112	3119	3126	3133	3141	3148	3155	1	1	2	3	4	4	5	6	7
	0	1	2	3	4	5	6	7	8	9	1	2	3	4	5	6	7	8	9

ANTI-LOGARITHMS

Proportional Parts

	0	1	2	3	4	5	6	7	8	9	1	2	3	4	5	6	7	8	9
·50	3162	3170	3177	3184	3192	3199	3206	3214	3221	3228	1	1	2	3	4	4	5	6	7
·51	3236	3243	3251	3258	3266	3273	3281	3289	3296	3304	1	2	2	3	4	5	5	6	7
·52	3311	3319	3327	3334	3342	3350	3357	3365	3373	3381	1	2	2	3	4	5	5	6	7
·53	3388	3396	3404	3412	3420	3428	3436	3443	3451	3459	1	2	2	3	4	5	6	6	7
·54	3467	3475	3483	3491	3499	3508	3516	3524	3532	3540	1	2	2	3	4	5	6	6	7
·55	3548	3556	3565	3573	3581	3589	3597	3606	3614	3622	1	2	2	3	4	5	6	7	7
·56	3631	3639	3648	3656	3664	3673	3681	3690	3698	3707	1	2	3	3	4	5	6	7	8
·57	3715	3724	3733	3741	3750	3758	3767	3776	3784	3793	1	2	3	3	4	5	6	7	8
·58	3802	3811	3819	3828	3837	3846	3855	3864	3873	3882	1	2	3	4	4	5	6	7	8
·59	3890	3899	3908	3917	3926	3936	3945	3954	3963	3972	1	2	3	4	5	5	6	7	8
·60	3981	3990	3999	4009	4018	4027	4036	4046	4055	4064	1	2	3	4	5	6	7	7	8
·61	4074	4083	4093	4102	4111	4121	4130	4140	4150	4159	1	2	3	4	5	6	7	8	9
·62	4169	4178	4188	4198	4207	4217	4227	4236	4246	4256	1	2	3	4	5	6	7	8	9
·63	4266	4276	4285	4295	4305	4315	4325	4335	4345	4355	1	2	3	4	5	6	7	8	9
·64	4365	4375	4385	4395	4406	4416	4426	4436	4446	4457	1	2	3	4	5	6	7	8	9
·65	4467	4477	4487	4498	4508	4519	4529	4539	4550	4560	1	2	3	4	5	6	7	8	9
·66	4571	4581	4592	4603	4613	4624	4634	4645	4656	4667	1	2	3	4	5	6	7	8	10
·67	4677	4688	4699	4710	4721	4732	4742	4753	4764	4775	1	2	3	4	5	7	8	9	10
·68	4786	4797	4808	4819	4831	4842	4853	4864	4875	4887	1	2	3	4	6	7	8	9	10
·69	4898	4909	4920	4932	4943	4955	4966	4977	4989	5000	1	2	3	5	6	7	8	9	10
·70	5012	5023	5035	5047	5058	5070	5082	5093	5105	5117	1	2	4	5	6	7	8	9	11
·71	5129	5140	5152	5164	5176	5188	5200	5212	5224	5236	1	2	4	5	6	7	8	10	11
·72	5248	5260	5272	5284	5297	5309	5321	5333	5346	5358	1	2	4	5	6	7	9	10	11
·73	5370	5383	5395	5408	5420	5433	5445	5458	5470	5483	1	3	4	5	6	8	9	10	11
·74	5495	5508	5521	5534	5546	5559	5572	5585	5598	5610	1	3	4	5	6	8	9	10	12
·75	5623	5636	5649	5662	5675	5689	5702	5715	5728	5741	1	3	4	5	7	8	9	10	12
·76	5754	5768	5781	5794	5808	5821	5834	5848	5861	5875	1	3	4	5	7	8	9	11	12
·77	5888	5902	5916	5929	5943	5957	5970	5984	5998	6012	1	3	4	6	7	8	10	11	12
·78	6026	6039	6053	6067	6081	6095	6109	6124	6138	6152	1	3	4	6	7	8	10	11	13
·79	6166	6180	6194	6209	6223	6237	6252	6266	6281	6295	1	3	4	6	7	9	10	12	13
·80	6310	6324	6339	6353	6368	6383	6397	6412	6427	6442	1	3	4	6	7	9	10	12	13
·81	6457	6471	6486	6501	6516	6531	6546	6561	6577	6592	2	3	5	6	8	9	11	12	14
·82	6607	6622	6637	6653	6668	6683	6699	6714	6730	6745	2	3	5	6	8	9	11	12	14
·83	6761	6776	6792	6808	6823	6839	6855	6871	6887	6902	2	3	5	6	8	9	11	13	14
·84	6918	6934	6950	6966	6982	6998	7015	7031	7047	7063	2	3	5	6	8	10	11	13	14
·85	7079	7096	7112	7129	7145	7161	7178	7194	7211	7228	2	3	5	7	8	10	12	13	15
·86	7244	7261	7278	7295	7311	7328	7345	7362	7379	7396	2	3	5	7	8	10	12	14	15
·87	7413	7430	7447	7464	7482	7499	7516	7534	7551	7568	2	3	5	7	9	10	12	14	16
·88	7586	7603	7621	7638	7656	7674	7691	7709	7727	7745	2	4	5	7	9	11	12	14	16
·89	7762	7780	7798	7816	7834	7852	7870	7889	7907	7925	2	4	5	7	9	11	13	14	16
·90	7943	7962	7980	7998	8017	8035	8054	8072	8091	8110	2	4	6	7	9	11	13	15	17
·91	8128	8147	8166	8185	8204	8222	8241	8260	8279	8299	2	4	6	8	10	11	13	15	17
·92	8318	8337	8356	8375	8395	8414	8433	8453	8472	8492	2	4	6	8	10	12	14	15	17
·93	8511	8531	8551	8570	8590	8610	8630	8650	8670	8690	2	4	6	8	10	12	14	16	18
·94	8710	8730	8750	8770	8790	8810	8831	8851	8872	8892	2	4	6	8	10	12	14	16	18
·95	8913	8933	8954	8974	8995	9016	9036	9057	9078	9099	2	4	6	8	10	12	14	17	19
·96	9120	9141	9162	9183	9204	9226	9247	9268	9290	9311	2	4	6	8	11	13	15	17	19
·97	9333	9354	9376	9397	9419	9441	9462	9484	9506	9528	2	4	7	9	11	13	15	17	20
·98	9550	9572	9594	9616	9638	9661	9683	9705	9727	9750	2	4	7	9	11	13	16	18	20
·99	9772	9795	9817	9840	9863	9886	9908	9931	9954	9977	2	5	7	9	11	14	16	18	21
	0	1	2	3	4	5	6	7	8	9	1	2	3	4	5	6	7	8	9